Biomedical Applications of Chemical Engineering

Topical Conference at the 2012 AIChE Annual Meeting

Pittsburgh, Pennsylvania, USA
28 October - 2 November 2012

ISBN: 978-1-62276-745-8

Printed from e-media with permission by:

Curran Associates, Inc.
57 Morehouse Lane
Red Hook, NY 12571

CURRAN ASSOCIATES INC. proceedings.com

Some format issues inherent in the e-media version may also appear in this print version.

Printed by Curran Associates, Inc. (2013)

For permission requests, please contact AIChE
at the address below.

AIChE
3 Park Avenue
New York, NY 10016-5991

Phone: (203) 702-7660
Fax: (203) 775-5177

www.aiche.org

Additional copies of this publication are available from:

Curran Associates, Inc.
57 Morehouse Lane
Red Hook, NY 12571 USA
Phone: 845-758-0400
Fax: 845-758-2634
Email: curran@proceedings.com
Web: www.proceedings.com

TABLE OF CONTENTS

(6a) Topical Plenary: Nanotechnology and Bioengineering in an Evolving Chemical Engineering World: The Next Generation of Recognitive, Intelligent Medical Microdevices

Nicholas A. Peppas, Dept. of Chemical Engineering, The University of Texas at Austin, Austin, TX

We will discuss first how the field of chemical engineering has contributed to major developments in bioengineering and nanotechnology, offering a historical prospective of the field and its leading scientists. Then, we will discuss how engineering the molecular design of intelligent biomaterials by controlling recognition and specificity is the first step in coordinating and duplicating complex biological and physiological processes. We will address design and synthesis characteristics of artificial molecular structures capable of specific molecular recognition of biological molecules. Recent developments in protein delivery have been directed towards the preparation of targeted formulations for protein delivery to specific sites, use of environmentally-responsive polymers to achieve pH- or temperature-triggered delivery, usually in modulated mode, and improvement of the behavior of their mucoadhesive behavior and cell recognition. Molecular imprinting and microimprinting techniques, which create stereo-specific three-dimensional binding cavities based on a biological compound of interest can lead to preparation of biomimetic materials for intelligent drug delivery, drug targeting, and tissue engineering. We have been successful in synthesizing novel glucose-binding molecules based on non-covalent directed interactions formed via molecular imprinting techniques within aqueous media.

(60a) Nanoparticle Cancer Therapeutics: Concept to Clinic

Mark E. Davis, Chemical Engineering, California Institute of Technology, Pasadena, CA

Nanoparticle-based therapeutics containing either small molecule anti-cancer agents (CRLX101 (formerly called IT-101); Phase II) or siRNA (CALAA-01; Phase I) are described. The key component in each nanoparticle is a cyclodextrin-containing polymer that endows it with multifunctional behavior. Relationships between nanoparticle design and function in animal models are illustrated. Results from preclinical animal models and human clinical trials are compared to illustrate similarities and differences between behaviors in animals and humans. For example, with CRLX101, features designed into the nanoparticles such as the ability to have long circulation times and long drug release in tumors cells has been observed in mice and humans. With CALAA-01, short circulation times observed in mice and non-human primates is also seen in humans. Additionally, the ability of CALAA-01 to release siRNA and have it perform RNAi inside of cancer cells within the tumors of animals and humans is verified using RACE analyses. While many of the features of the nanoparticles do translate from animals to humans, there are issues that do arise when treating humans. Some of these issues will be discussed.

(60b) Multi-Reservoir Drug Therapies...From Academic Lab to Startup Company to Clinical Demonstration

John T. Santini Jr., On Demand Therapeutics, Inc., Tyngsboro, MA

As pharmaceutical drugs become more potent, targeted, and complex, the importance of systems designed to maintain drug stability and precisely control the delivery of those drugs to the body increases. Therefore, a real need exists for novel materials and devices that can: (1) store one or more drugs, (2) protect them from the body until they are needed, and (3) controllably release the drugs at the desired time. Advances in microfabrication and hermetic sealing technologies have enabled the creation of multi-reservoir drug delivery systems that address these needs. Precise control over the release of drug from each reservoir is enabled by novel formulation techniques, the application of outside energy sources such as lasers, and/or the integration of pre-programmed microprocessors, wireless telemetry, or biosensors.

Startup companies have proven to be an effective way to quickly translate scientific and technological discoveries from academic labs into new medical products that benefit patients. This talk will include a discussion of: (1) the development and first clinical demonstration of an implantable, wireless, multi-reservoir drug therapy for osteoporosis, and (2) the various issues impacting the successful translation of academic science to the clinic.

(60d) Transdermal Drug Delivery: Translation From Chemical Engineering Laboratories Into the Clinic

Samir Mitragotri, Chemical Engineering, University of California, Santa Barbara, Santa Barbara, CA

Transdermal drug delivery provides an advantageous alternative to injections for delivering drugs into the body. It avoids the pain and needle—phobia associated with injections while providing sustained release of drugs. However, delivery of drugs into skin is limited by the high barrier properties of skin owing largely to its topmost layer, the stratum corneum. Accordingly, applications of transdermal delivery are limited to a handful of small drugs. Several technologies have been developed to enhance skip permeability to drugs after topical application. These technologies utilize physical and chemical means to reversibly permeabilize stratum corneum. These technologies have yielded devices as well as formulations that can be applied on the skin for local and systemic delivery of drugs. My presentation will present an overview of these technologies with an emphasis on their transition into the clinic. Applications of these technologies in the clinical setting and clinical problems addressed by these technologies will be discussed.

(60e) New Approaches to Treating Brain Tumors

Gary Gallia, Neurosurgery, Johns Hopkins Univ. SOM, Baltimore, MD, Betty Tyler, Neurosurgery, Johns Hopkins University SOM, Baltimore, MD and Henry Brem, Neurosurgery, Oncology and Biomedical Engineering, Johns Hopkins University, Baltimore, MD

New Approaches to Treating Brain Tumors

Gary Gallia, Betty Tyler, and Henry Brem

Departments of Neurosurgery, Oncology, and Biomedical Engineering

Johns Hopkins University, Baltimore, Maryland

The technology of local delivery is being utilized to target multiple pathways for brain tumor therapy. We will review the challenges of developing Gliadel, the first polymer drug delivery system for brain tumors. Through combination of surgical resection, Gliadel, radiation therapy and oral temozolomide (TMZ), median patient survival for brain tumor patients has increased from 9 to 21 months. Recent clinical trials with resistance modifiers and trials with Gliadel for brain metastases have also led to improved survival. To further increase survival we have been exploring the local delivery of temozolomide and have shown that in animal glioma models we can achieve 37.5% long term survival. We have also incorporated paclitaxel, a mitotic inhibitor, into a thermo-sensitive gel depot which provides local delivery, enhances efficacy, and limits systemic toxicity. In combination with TMZ and radiation therapy this depot significantly increases long term survivors and improves median survival compared to either therapy alone. This data led to a Phase I clinical trial. Rapamycin, an MTOR inhibitor, anti-angiogenic agent and anti-proliferative agent, currently used for coronary stents, has been efficiently incorporated and intracranially delivered using biodegradable controlled release beads. We demonstrated its safety for CNS delivery as well as it significantly prolonged survival in our 9L rodent glioma model. Other anti-angiogenic agents are being explored as well, including Fc-endostatin, a recombinant human endostatin conjugated to the Fc domain of IgG, which has shown significantly prolonged survival. Minocycline, a semi-synthetic tetracycline, has been shown to be an effective adjuvant therapy when combined with either radiation therapy or oral temozolomide, two common methods of treating glioblastoma. Direct intracranial delivery using microchips has been a focus of our laboratory in collaboration with colleagues at MIT. Both biodegradable "passive" chips as well as microelectromechanical systems, or "active" chips, have shown efficacy in delivering chemotherapeutic agents for the treatment of both glioma as well as metastatic cancer models. Intracranial microdialysis has proven to be a very useful technique to determine real time intracranial concentrations of systemically delivered drugs. This helps to determine if effective concentrations are being reached at the site of the brain tumor. To date we have contributed to our overall objective of developing better technologies for treating brain tumor patients using locally delivered therapies.

(89a) Practical Issues Encountered When Attempting to Field New Technology to Soldiers

Keith LeJeune, FLIR Systems, Export, PA

As technologists and inventors, much of our work focuses on gaining a clearer understanding of the fundamentals of science and establishing the potential application of new technologies, processes, and/or methods. Within the context of defense against weapons of mass destruction, much of the applications research searches for answers to questions like "What is the capacity of a decontaminant?" or "What is the limit of detection for a particular sensor technology?". These are very important questions / issues to address as they identify the potential of a particular technology.

I have spent the last ten years working to transition bio-based technologies for chemical and biological defense to the user community. Most of the initial work focused on answering similar basic technology application questions. More recently, the focus has been on different questions. "How do you package a sensor to keep it from breaking during transport"? "How can a soldier wearing protective gear and 60 minutes of air in an SCBA system transport this sprayer to an event sight"? In hindsight, having a better appreciation for the issues encountered in the past few years would have significantly impacted the approach taken earlier. This presentation will describe some lessons-learned throughout this process. I gained this awareness through experience, which unfortunately costs time and money to acquire. The CBRE threat mitigation community would significantly benefit from the development of programs that provide technology developers with greater access/visibility to the constraints on the warfighter.

(89b) Combatting Weapons of Mass Destruction Through Innovation in Convergence of Chemical Engineering and Biology

Stephen Lee, US Army Research Office, Research Triangle Park, NC

Weapons of Mass Destruction (WMD) continue to constitute a threat overhanging society with the ever-expanding capability to more readily prepare some of the core WMD materials including chemical and biological warfare agents. Critical needs continue to detection of, protection against, and destruction of these threats. The convergence of Biology and Chemical Engineering is beginning to elucidate exciting examples where advanced, multi-functional concepts, combining detection, protection, and destruction in a single system, are being developed. Dynamic and responsive properties are known throughout nature and more recently have been realized in designing non-biological systems. There are many examples of responsive nanomaterials such as abiotic enzymes, responsive micelles, electroactive polymers, and responsive liquid crystalline surfaces to name a few. These materials all change morpohology, structure, or function in response to an external stimulus such as pH, oxidation, conductivity, or a chemical trigger. Key fundamental issues around chemical reactivity, mass transport, nanostructured design, biological/abiological interfaces, and biomimetic systems are all critical for advancement of these systems. A challenge for chemical engineering is to use the chemistry and designs of biology in the dynamic systems to develop complete multi-functional solutions , which react with the threat to trigger a reaction that indicates the presence of the threat, neutralizes the threat, and then indicates the threat has been neutralized. Advances in these areas will ultimately revolutionize smart systems for protection from WMDs.

(89c) Preparing Against Biological and Chemical Threats: A Strategic Focus in an Unpredictable Future

Alan Rudolph, Joint Science and Technology Office, Defense Threat Reduction Agency, FORT BELVOIR, VA

Perhaps one of the biggest challenges of our generation is preparing against health and material threats to our nation. Considerable policy attention to global health security has defined these challenges in the context of both threats from adversarial intent as well as safety and security associated with accidental or pandemic events that can occur at any time and rapidly move a potentially catastrophic event to our shores from anywhere on the planet. The mission of the Joint Science and Technology Office at the Defense Threat Reduction Agency is to invest in innovative programs that will best prepare our nation and our defenses against such threats. The challenge we face is clear. With the widening aperture of the threat space, how do we create a strong position from which to respond? We have created a strategic plan that recognizes the dynamic nature of the threat and the need for technology to be able to pivot quickly as time is the most precious of response resources. The time continuum of response recognizes a need for advanced warning and rapid countermeasures following the alert. In this context, we have crafted a strategic plan that recognizes the need for new diagnostic platforms that are highly distributed ("point of need") and interoperable and connected to medical and public health reporting systems. The biological and chemical surveillance ecosystem we seek is intended to provide time to reduce untoward consequences of these threats. The ability to rapidly mount a countermeasure response is another focus on our strategic plan in which the realization of rapid manufacturing and validation of new medical countermeasures and physical protection systems offers the best opportunity to both mitigate the threat and return our society to equilibrium. New investments in each of these focused strategic areas will be presented.

(89e) Engineering a New Rapid-Response Vaccine Capability

Anton P. J. Middelberg, Australian Institute for Bioengineering and Nanotechnology, The University of Queensland, Brisbane QLD 4072, Australia

In chemical engineering, the concept of a timely response to disturbance is taught in undergraduate courses. A complex system subject to disturbance is controlled by actions that mitigate the impact of disturbance before that disturbance magnifies to disrupt the entire system. The time constant of response must be smaller than the time constant over which the disturbance will amplify. In vaccine engineering, this is not presently the case. The disturbance (e.g. a virus) can propagate faster than we can mount a vaccine response. This fact was amply demonstrated by the (H1N1) 2009 pandemic which led to social disruption and strain on the health system, even though the disease was mild in most cases. Also central in chemical engineering is the concept of a product made to a cost the market can afford. Although vaccination has led to the eradication of smallpox, around 15 million people still die annually from infectious disease, with more than half of them being children under the age of 5. A new vaccine engineering paradigm is urgently needed. Here we report a new manufacturing approach which starts with non-infectious protein and builds structure, and hence antigenicity, without building infectivity. Our approach relies on the expression and purification of proteins prepared in bacterial cell factories which are then processed *in vitro* into virus-like particles (VLPs). Protein production in highly efficient bacteria maximises speed and efficiency while also minimising the need for high-level biocontainment during manufacture. The approach is both exciting and challenging, raising a number of research questions amenable to solution by chemical engineers. Combined, these aspects have the potential to move us toward a vaccine future based on fundamental chemical and molecular engineering, using fast, scaleable and cheap technological approaches.

(104a) Protein Analogous Micelles: Versatile, Modular Nanoparticles

Matthew Tirrell, Institute for Molecular Engineering, University of Chicago, Chicago, IL

Peptides are functional modules of protein macromolecules that can be displayed apart from the whole protein to create biofunctional surfaces and interfaces, or can be re-assembled in new ways to create synthetic mimics of protein structures. Each of these routes are being employed to gain new insight into protein folding and to develop new, functional, biomolecular materials. Examples of work from our laboratory in this area using peptide-lipid conjugate molecules (peptide amphiphiles) will be discussed relating to multi-functional surfaces, DNA-binding peptide assemblies, synthetic vaccines, and protein analogous micelles for cancer and cardiovascular therapeutics. Spherical micelles with characteristic dimensions of less than 20 nanometers are very effective in avoiding clearnace before reaching targeted pathlogical tissues. Particular applications to targeting vulnerable atherosclerotic plaque and solid tumors will be described. Aspects of structure-function relationships, stability and control of geometry of these assemblies will be discussed. Protein anaologous micelles represent the most flexible, potent and versatile class of nanoparticles for any applications.

(104b) Molecular Transport and Structure of Surface Modified Nanopores:

Igal Szleifer, Biomedical Engineering, Northwestern University, Evanston, IL

Natural and man-made pores and channels of nanoscale dimensions display unique ionic and protein transport behavior. Nanopores modified with supramolecular chemical species (such as polyelectrolyte brushes) have dimensions that are similar to the range of the electrostatic interactions, and also to the molecular size of the tethered macromolecules. In cells, Nuclear Pore Complexes (NPC) control the transport of species between the cytoplasm and the nucleus using disordered proteins as gate keepers. The competition between molecular and interaction length scales, as well as the geometry of the surfaces, creates interesting possibilities for the creation of stimuli responsive gates and ion channels and for the fundamental understanding of the interplay between molecular organization and charge transport in nanoconfined environments. In this talk we will discuss recent theoretical developments that enable us to study the structural properties of the modified pores and the coupling that exists between the protonation state of the polyelectrolytes and ionic transport through them. The predictions from the theory are in excellent agreement with experimental observation and they provide a direct understanding of the important coupling that exists between protonation state, molecular organization and transport. We will demonstrate how the molecular organization within the pore is a strong function of the pore geometry and how these effects translate into the transport behavior. We will show a novel non-equilibrium approach to treat transport of ions and discuss the intrinsic coupling that exists between macromolecular organization within the nanopore and ion transport. Finally, we will show preliminary calculations of the spatial distribution of amino acids when the full NPC disorder proteins is taken into account. The importance of considering explicit molecular details of the surface modifiers as well as the coupling between chemical equilibrium, physical interactions and molecular organization will be discussed in detail.

(104c) The Micromechanics and Physics of Cancerous Cells Revealed Using the Tools of Nanoscience

Bartosz Grzybowski, Chemical and Biological Engineering, Northwestern University, Evanston, IL

The mainstream research in cancer biology focuses on the genetic and molecular signatures underlying the disease. An alternative and perhaps more physics-oriented approach is to study cancer from a perspective of an entire cell treated as a complex system characterized by a set of micromechanical properties and cell-motility characteristics. In my lecture, I will talk about both of these aspects. In the context of micromechanics, I will first describe a new cell-tomography technique which allows for the reconstruction of 3D shapes of cells subject to well-defined, geometrically regular microconfinements. The reconstructed shapes allow for the derivation of an analytical micromechanics model according to which cell's mechanical properties are determined by the deformability of the cell membrane and cortical actin and by the limited compressibility of the nucleus. This model does not involve any arbitrary parameters and is universal in the sense that it reproduces the shapes, volumes and surface areas of cells of different types under different experimental conditions. Unfortunately, analysis of cancerous vs. non-cancerous cells indicates little – if any – differences in their micromechanical properties. Thus, in search for the features distinguishing malignant from non-malignant cells the second part of the talk will focus on the ways in which cells move – again, over well defined microenvironments allowing for quantitative analyses. As I will show, non-cancerous and non-metastatic cells migrate in ways fundamentally different from metastatic cells. While non-metastatic cells are simple, diffusive random walkers, their non-metastatic variants employ space-searching strategies used by animal predators. In this spirit, metastatic "cellular predators" can be viewed as navigating optimally through human body in search for suitable conditions to thrive (as in the "seed and soil"). Importantly, in doing so they desynchronize their cytoskeletal dynamics and "lose" a characteristic time scale present in non-metastatic cells. These observations prompt a hypothesis that cancer metastasis entails "desynchronization" of a cytoskeletal "clock".

(104d) Artificial Viruses

Nicholas Kotov, Chemical Engineering, University of Michigan, Ann Arbor, MI

Self-organization of nanoparticles (NPs) and nanoscale objects represents one of the most dynamic areas of materials science. Better understanding of these phenomena is important from both fundamental and practical perspectives because nanoparticle self-organization processes

(1) identify similarities between biological and non-biological nanoscale species;

(2) lead to unusual optical properties from different combinations of nano- and microscale features;

(3) can potentially simplify manufacturing of electronic, photonic, and sensing devices.

In this presentation, the analogies between self-organized nanoparticle assemblies and viruses will be discussed and analyzed. We shall start with their thermodynamics and continue with the preparation of different NP assemblies resembling the structure and geometry of viruses. The individual NPs in the virus-like superstructures will be treated as an analog of a protein. Dynamic reorganization of NP superstructures will also be considered. The presentation will be concluded with parallels in the interactions of NP assemblies with cells and their utilization for drug delivery.

Relevant References.

1. Z. Tang, N. A. Kotov, M. Giersig, *Science,* **2002**, *297,* 237-240.
2. Tang, Z. Zhang, Z.; Wang Y.; Glotzer, S. C. Kotov, N. A., *Science*, **2006**, 314 (5797) 274-278.
3. J. Lee, A. O. Govorov, N. A. Kotov, *Angew. Chem. Intern. Ed.* **2005**, *44,* 7439-7442.
4. J. Lee, A. O. Govorov, N. A. Kotov, *Nature Materials.* **2007**, 6(4), 291-295.
5. S. Srivastava, A. Santos, K. Critchley, K.-S. Kim, P. Podsiadlo, K. Sun, J. Lee, C. Xu, G. D. Lilly, S. C. Glotzer, and N. A. Kotov, *Science*, **2010**, 327, 1355-1359.
6. M. Yang, K. Sun, N. A. Kotov, *J. Am. Chem. Soc.*, **2010**, *132* (6), pp 1860–187.
7. Y. Zhou, M. Yang, K. Sun, Z. Tang and N. A. Kotov, *J. Am. Chem. Soc.*, **2010**, *132* (17), 6006–6013
8. N. A. Kotov, Inorganic Nanoparticles as Protein Mimics, *Science,* **2010**, 330 (6001), 188-189.
9. S. I.Yoo, N.A. Kotov, *Angewandte Chemie*, **2011,** 50(22), 5110–5115.
10. Y. Xia et al, *Nature Nanotechnology*, **2011**, 6, 580-587.

(104e) Confinement Protection Effects of Mesoporous Silica SBA-15 On Myoglobin, in an Environment Inspired by the Groel/ES Chaperonin System

Michele Lynch[1], Justin Siefker[2] and Marc-Olivier Coppens[1], (1)Chemical & Biological Engineering, Rensselaer Polytechnic Institute, Troy, NY, (2)Dept. Chemical & Biological Engineering, Rensselaer Polytechnic Institute, Troy, NY

With the increasing prevalence of protein therapy drugs, such as antibodies, it is important to stabilize and protect these proteins in the body. The structure of proteins is more delicate than that of small-molecule drugs; e.g., many proteins denature and become inactive when exposed to proteases or extreme pH [1]. Microscopic, mesoporous silica SBA-15 particles are interesting candidates for drug delivery and sensing applications. SBA-15 contains a hexagonally ordered array of cylindrical pores. These pores have certain commonalities with the GroEL/ES complex, a pair of chaperone proteins found in bacteria cells that assist in the folding of new proteins [3]. More in particular, the pores of SBA-15 have a hydrophilic core, similar to the GroEL/ES complex, and exhibit similar electrostatics with respect to protein adsorption. We report on the denaturation protection from exposure to proteases and extreme pH, conferred to proteins adsorbed on the interior pore surface of SBA-15.

To test for the protective effects of SBA-15, myoglobin is adsorbed on the internal pore surface of SBA-15 particles, and these composites are exposed to the protease pepsin. Both free myoglobin and myoglobin/SBA-15 composites are suspended in phosphate buffer with a pH of 4.6. This pH was chosen because pepsin is only active in environments with a pH lower than 6.5, and optimally at a pH of 2.0 [2]. The activity of free and confined myoglobin is tested via a peroxidase enzyme activity assay, which is measured by UV-Vis spectroscopy. Activity measurements are normalized to those of free myoglobin in buffer with a pH of 7.2 (myoglobin's isoelectric point), and to the measured amount of adsorbed myoglobin. The myoglobin loading of all composites is below 50 mg/g SBA-15, this is within the Henry adsorption regime.

Our previous work had shown that adsorbing myoglobin to SBA-15 at its isoelectric point (pH 7.2) leads to an extraordinary increase in activity, around 700% [1]. However, at pH 4.6, the activity of confined myoglobin is reduced with respect to that of free myoglobin at the same pH. When free myoglobin is digested by active pepsin, its activity is reduced substantially. In contrast, confined myoglobin shows a nearly imperceptible drop in activity when exposed to pepsin. This result is encouraging, as it suggests that confinement protects the myoglobin from proteolysis.

Changes in secondary structure are observed with Fourier transform infrared spectroscopy (FTIR). Also the effect on myoglobin catalytic activity of restoring the pH of all samples to 7.2 after a temporary decrease in pH will be reported. This experiment is inspired by the behavior of the GroEL/ES complex, which can also refold denatured proteins. The results of this study elucidate the effects of confinement in SBA-15 on myoglobin with respect to extreme pH exposure.

[1] L.C. Sang and M.-O. Coppens (2011), *Phys. Chem. Chem. Phys.* **13**, 6689.

(104f) Inhibition of Viral Infection by Molecularly Imprinted Nanoparticles - A Synthetic Approach to Antiviral Therapy

Yen Wah Tong, Department of Chemical and Biomolecular Engineering, National University of Singapore, Singapore, Singapore and Niranjani Sankarakumar, Chemical and Biomolecular Engineering, National University of Singapore, Singapore, Singapore

Despite the current advances in the field of medicine and pharmaceutics, virus transmitted diseases are still one of the major health issues today with some of the most deadly and perplexing viruses being Ebola, HIV and SARS. Unfortunately there are no cure nor viable vaccines for all types of viral infections while currently existing therapies lose their effectiveness with time or due to viral mutations. To cater to the mounting demand for new antiviral strategies, we herein propose the use of high affinity polymeric receptors prepared by molecular imprinting technique to "catch" viruses.

Molecular imprinting is an effective technique that can generate tailor-made artificial materials possessing comparable molecular recognition and binding capabilities as with natural biomolecules like antibodies. It involves the creation of three dimensional highly specific binding sites in a polymer matrix ('the lock') that are complementary to the size, shape and functional group orientation of the 'substrate' molecule ('the key'). Thus, molecularly imprinted polymers (MIPs) have been extensively used as substitutes for antibodies or enzymes in a wide range of technologies such as separation and purification, sensing, catalysis and drug delivery. In this research, our aim is to create surface imprints of viruses on polymeric nanoparticles and to employ the particles for "capture" and removal of any active, disease-causing viruses.

Virus imprinted polymeric nanoparticles (vMIPs) were fabricated using different functional monomers and surface imprinting strategies with fr bacteriophages employed as a model virus. The imprinted polymeric particles prepared were of sizes ranging from 40 nm to 570 nm using methyl methacrylate and ethylene glycol dimethacrylate as the functional and cross-linker monomers, respectively. A significant reduction in phage concentrations were observed after just 3 h of exposure to vMIPs. The imprinted particles further displayed remarkable positive anti-viral effect that significantly hindered phage infection of host bacteria as compared to the controls. To the best of our knowledge, this is the first report demonstrating the use of molecularly imprinted polymers as anti-viral agents targeted at infectious disease treatment.

The synthesis of selenium nanoparticles via a simple fast reaction and its antibacterial application

Qi Wang[1] *and Thomas J. Webster*[1,2]
[1]*Bioengineering Program and* [2]*Department of Chemical Engineering, College of Engineering, Northeastern University, Boston, MA 02115, U.S.A.*

Abstract

In this study, selenium nanoparticles were synthesized and coated on materials using a simple fast precipitation method. The amount, roughness, adhesion and other properties of the nano-sized coatings were characterized. By coating with selenium nanoparticles, the materials, specifically polycarbonate and paper towels, were introduced with antibacterial properties inhibiting bacterial growth. The objective of the present study was to coat selenium nanoparticles on materials and test their effectiveness preventing bacterial growth. The results showed that the selenium nanoparticles coated on the materials were well distributed semi spherical and 50-100nm in diameter. Most importantly, selenium coatings significantly inhibited *Staphylococcus aureus* growth to 27% and 10% after 72 hours on polycarbonate and paper towels, respectively, compared to the uncoated materials. Therefore, this study suggested that coating materials with selenium nanoparticles is a fast and effective way to reduce bacterial infections, which could potentially used for various materials.

Introduction

Bacterial infections are commonly found in hospital environment, such as forming biofilms on various medical devices, bacteria attachment on the wall paper of the clinic room resulting in cross contamination during flu seasons and contamination of hands caused by paper towels hanging in sink splash zones.[1-4] *S. aureus* (*Staphylococcus aureus*) is one of the bacterium commonly found in numerous infections. Each year, there are 11 million outpatient/emergency room visits and 464,000 hospital admissions in the U.S. alone due to *S. aureus* infections.[5] *S. aureus* biofilms have been found on a wide range of medical devices including prosthetic heart valves, central venous catheters, orthopedic prostheses, contact lenses, endocarditis, otitis media and osteomyelitis.[6] Therefore, it is significant to develop a method to prevent bacterial infections on materials.

Different from plenty of nano-sized metal-based antibacterial materials, such as zinc oxide nanoparticles,[7] silver nanoparticles,[8-10] copper and its alloys,[11-12] that may cause potential toxicity to mammalian cells,[13-15] selenium is a naturally occurring micronutrient, which is fundamental for 25 selenoproteins or enzymes in the human body with selenocysteine at their active center.[16] Therefore, in this study, selenium nanoparticles were coated on polycarbonate and paper towel surfaces through a quick precipitation method and the properties of selenium coated paper were characterized. In addition, such materials were tested for their effectiveness to kill bacteria and/or prevent bacteria from attaching in the bacterial assays. The results showed that the selenium coatings successfully introduced antibacterial properties to polycarbonate and paper towels, revealing a promising nano-selenium based method to prevent bacterial infections on various materials.

Materials and Methods

Selenium nanoparticles were synthesized through a simple reaction between glutathione and sodium selenite (4:1 molar mixture) and were precipitated on the surface of polycarbonate (McMASTER-CARR, 85585k82, 0.13mm thickness, cut into round films 7.01mm in diameter) and paper towels (Tork Advanced, MB550A Hand Towel, cut into round chips, 7.01mm in diameter) for either 30s or 60s. Sodium hydroxide was added to bring the pH of the solution to the alkaline regimen to start the reaction. As the size of selenium nanoparticles on the surface is influenced by the concentration of NaOH and the coating time, two coating conditions were used in this study. One was using a 0.5M NaOH solution for 30s and the other one was using a 1.0M NaOH solution for 60s. After coating, tape tests (D3359-09e2: Standard Test Methods for Measuring Adhesion by Tape Test) were used to test the strength of adhesion of the selenium nanoparticles on the substrate surfaces. AAS (Atomic Absorption Spectroscopy) was used to measure the amount of coated selenium on polycarbonate surfaces. The SEM (Scanning Electron Microscope, HITACHI 2700) images of the substrate surfaces were taken to determine the size, the distribution and coverage of selenium nanoparticles. AFM (Atomic Force Microscope, MFP3D, Asylum Research, Sharp tipped cantilever, K = 0.06N/M, Contact Mode) was used to demonstrate that the coated selenium nanoparticles increased the surface selenium exposure and surface roughness of polycarbonate.

In order to test the effectiveness of selenium coatings inhibiting biofilm formation, bacteria experiments were implemented. A bacteria cell line of *Staphylococcus aureus* was obtained in freeze-dried form from the American Type Culture Collection (catalog number 25923). Selenium coated samples were rinsed with 75% ethanol for 20 minutes for sterilization purposes and were left in the sterile petri dishes for 30 minutes to completely dry. Then, the samples were transferred to a 24-well plate, treated with the prepared bacterial solutions (10^6 bacteria/ml, *Staphylococcus aureus*) and cultured for either 24, 48 or 72 hours in an incubator (37°C, humidified, 5% CO_2). After the treatment, the samples were rinsed with a PBS (phosphate buffered saline) solution twice and placed into 1.5ml microfuge tubes with 1ml of PBS. These tubes were shaken at 3000 rmp for 15 minutes on a vortex mixer to release the bacteria attached on the surface into the solution. Solutions with bacteria were spread on agar plates and bacteria colonies were counted after 18 hours of incubation. Bacterial tests were conducted in triplicate and repeated three times. Data were collected and the significant differences were assessed with the probability associated with a one-tailed Student's t-test. Statistical analyses were performed using Microsoft Excel (Redmond, WA).

Results and Discussion

Most of the nanoparticles were approximately 50-100nm in diameter and well distributed on the polycarbonate (Figure 1) and paper towel (Figure 2) surfaces. On polycarbonate surfaces, selenium nanoparticles coated with a condition of 1.0M NaOH for 60s were larger than those with the condition of 0.5M NaOH for 30s. The concentration of nanoparticles on the polycarbonate surfaces were on average $19.34 g/m^2$ and $20.95 g/m^2$ for the 0.5M NaOH for 30s condition and for the 1.0M NaOH for 60s condition, respectively. While on paper towels, the concentration of of selenium nanoparticles on the coated paper towel surface was $69.00\ g/m^2$, which was 4 times larger than the amount on polycarbonate under the same coating conditions. One of the reason was that the fibrous structure of the paper towel allowed the samples to hold more selenium nanoparticles. Tape tests removed about 50% and 75% of the selenium nanoparticles from polycarbonate surfaces for the 0.5M NaOH for 30s condition and for the 1.0M NaOH for 60s condition, respectively, according to AAS results. This result indicated that selenium nanoparticles with larger sizes might have less adhesion to an underlying substrate than smaller selenium nanoparticles.

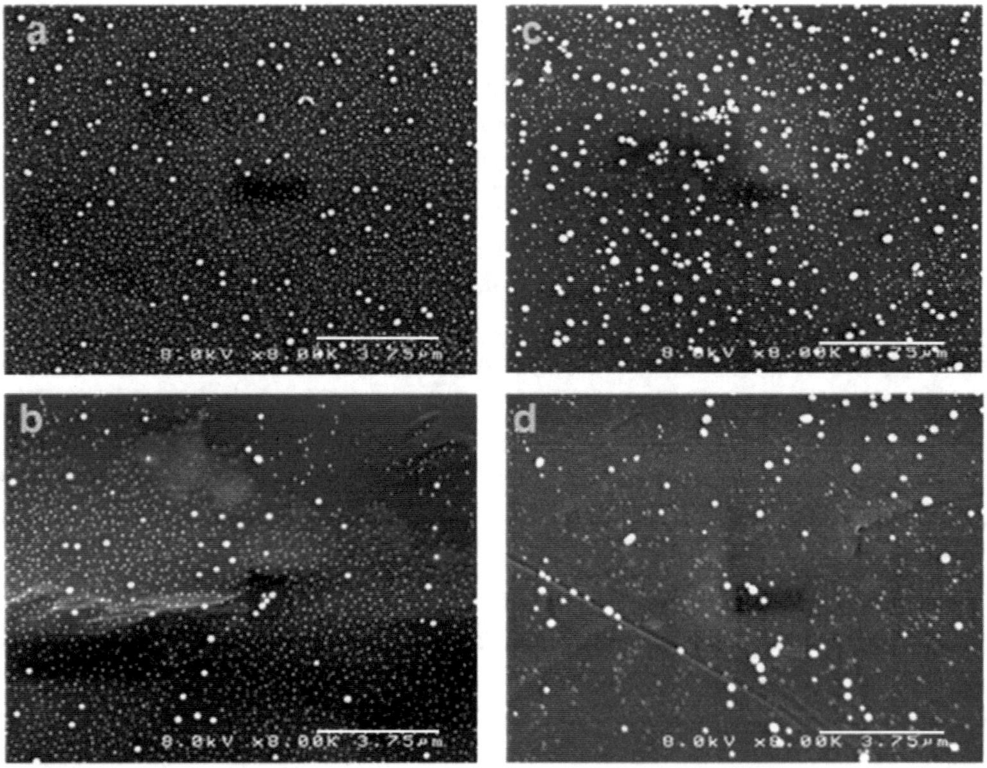

Figure 1. SEM images of selenium coated polycarbonate samples before and after a tape test. (a) 0.5M NaOH, coating for 30 seconds and before a tape test. (b) 0.5M NaOH, coating for 30 seconds and after a tape test. (c) 1.0M NaOH, coating for 60 seconds and before a tape test. (d) 1.0M NaOH, coating for 60 seconds and after a tape test. The coverage of selenium nanoparticles on the surface for image (a), (b), (c) , (d) are 11.74%, 6.94%, 11.38%, 5.88% respectively.

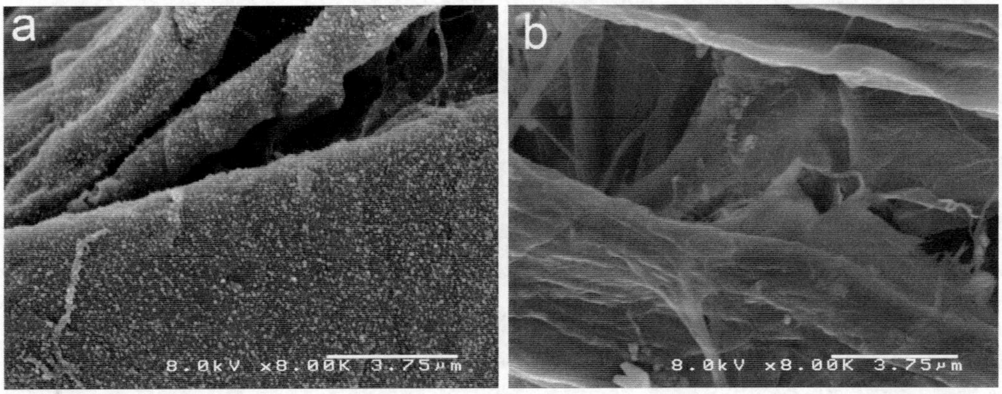

Figure 2. SEM images of selenium coated (image a) and uncoated paper (image b) towel samples. The coating condition for image a was 0.5M NaOH for 30 seconds. The concentrations of selenium on paper towels from AAS measurement were 69.00 g/m^2 for selenium coated paper towels and 0 g/m^2 for uncoated paper towels.

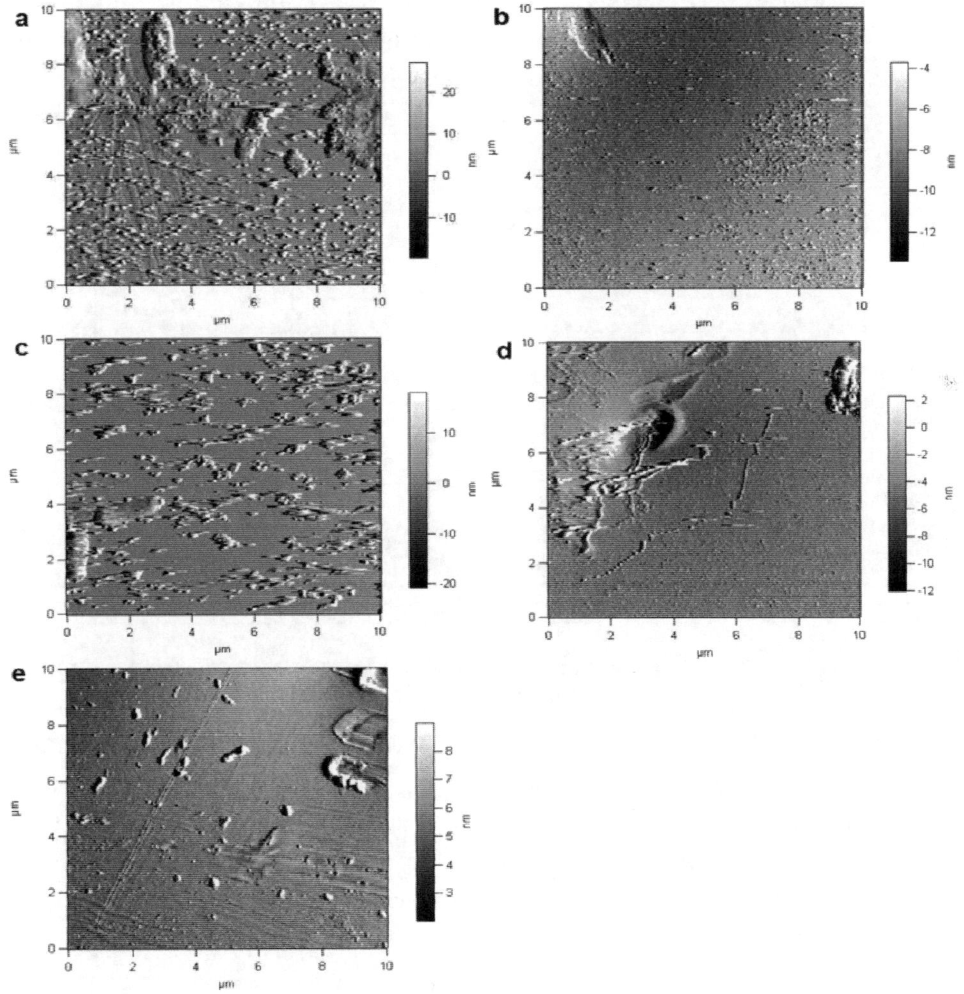

Figure 3. AFM images and RMS (root mean square, scan area = 10um×10um) roughness of

polycarbonate surface. (a) Selenium coated with 0.5M NaOH for 30s, RMS roughness = 45.997nm. (b) Selenium coated with 0.5M NaOH for 30s, after tape test, RMS roughness = 21.731nm. (c) Selenium coated with 1.0M NaOH for 60s, RMS roughness = 53.084nm. (d) Selenium coated with 1.0M NaOH for 60s, after tape test, RMS roughness = 34.925nm. (e)Uncoated sample, RMS roughness = 14.898nm.

From the AFM images (Figure 3), The RMS (root mean square) roughness for the selenium coated surface at a condition of 0.5M NaOH for 30s and at a condition of 1.0M NaOH for 60s were 45.997nm and 53.084nm, respectively. After the tape test, the RMS roughness decreased to 21.731nm and 34.925nm for the these two coating conditions. The RMS roughness of the uncoated surface was 14.898nm, which was much smaller than the selenium coated polycarbonate samples. So, there was a significant increase in roughness, therefore surface area, after coating with selenium nanoparticles.

Experiments with bacteria (specifically, *Staphylococcus aureus*) showed high effectiveness for the selenium coated materials at killing bacteria or preventing bacteria from attaching compared with the uncoated surfaces (Figure 4 and Figure 5). Compared to uncoated samples, more than 91%, 15% and 73% of the bacteria (compared with uncoated polycarbonate) died or had been removed after 24, 48 and 72 hours on the selenium coated polycarbonate. The selenium coated paper towels had 88.6%, 88.9% and 88.8% less bacteria attached than the uncoated paper towels after treatment with the bacteria solution for 24, 48 and 72 hours, respectively. Importantly, this was accomplished without using antibiotics.

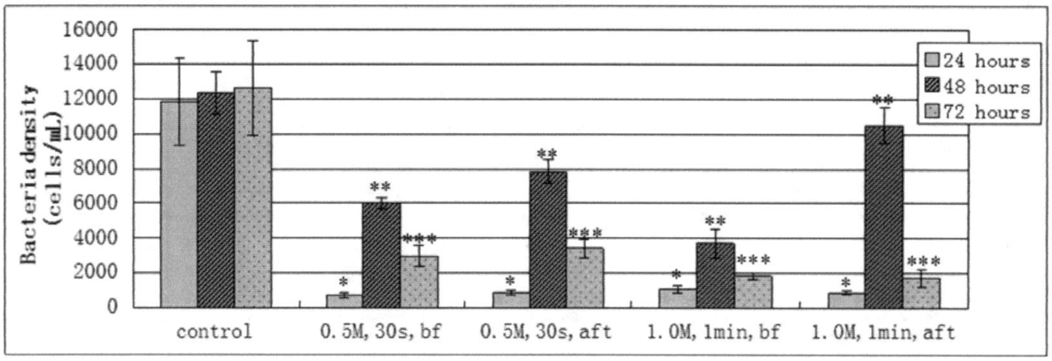

Figure 4. Bacteria (*S. aureus*) growth on the surface of polycabonate. Polycarbonate samples were treated with bacteria (*S. aureus*) in 0.03% TSB (Tryptic Soy Broth) and were incubated for 24, 48 or 72 hours. The media was changed with 0.03% TSB every 24 hours for those samples incubated at 48 hours or 72 hours. The control group is uncoated polycarbonate. bf = before tape test; aft = after tape test. Data=Mean ± standard deviation by mean, n=3; *$p<0.05$ compared with control group (uncoated polycarbonate) after 24 hours; **$p<0.004$ compared with control group after 48 hours; ***$p<0.02$ compared with control group after 72 hours.

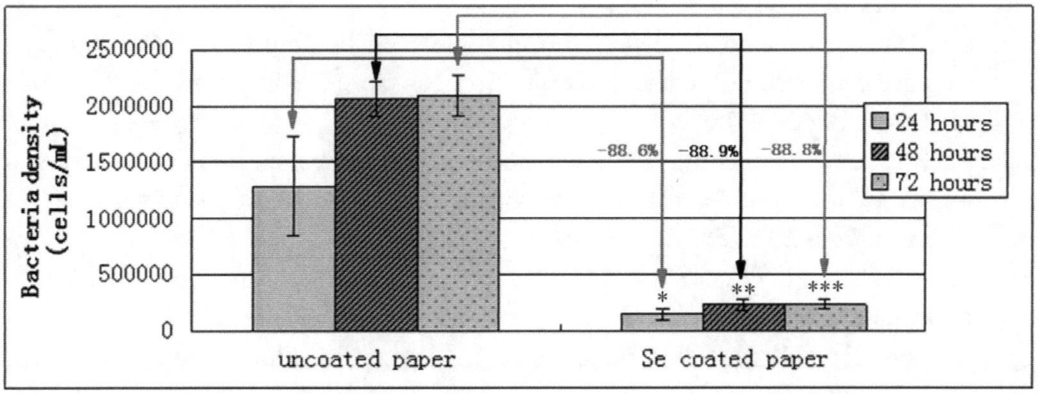

Figure 5. Bacteria (*S. aureus*) growth on the surface of uncoated selenium coated paper towels. Paper towel samples were treated with bacteria (*S. aureus*) in 0.03% TSB (Tryptic Soy Broth) and were incubated for 24, 48 or 72 hours. The control group is the uncoated paper towels. Data=Mean ± standard deviation, n=3; *p<0.02 compared with control group (uncoated paper) after 24 hours; **p<0.002 compared with the control group after 48 hours; ***p<0.002 compared with control group after 72 hours.

Conclusions

Selenium nanoparticles were synthesized and coated on polycarbonate and paper towels by a simple fast reaction occurring in 60s. The coating of selenium nanoparticles on polycarbonate and paper towels introduced significant and effective anti-bacterial properties to the material surfaces which deserves further investigation.

Acknowledgements

The authors thank Dr. Justin Seil for help with the bacteria experiments and Dr. Vera Fonseca for help with the AFM images. Funding from the Hermann Foundation is also acknowledged.

References

1. Costerton JW, Z. Lewandowski, Caldwell DE, Korber DR, Lappin-Scott HM. Microbial biofilms. *Annu. Rev. Microbiol.* 1995; 49:711–745.
2. Zottola EA., and Sasahara KC. Microbial biofilms in the food processing industry—should they be a concern. *Int. J. Food Microbiol.* 1994; 23: 125–148.
3. Darouiche RO, Treatment of infections associated with surgical implants. *N. Engl. J. Med.* 2004; 350:1422–1429.
4. Hattula JL, Steven PE. A descriptive study of the handwashing environment in a long-term

care facility. Clin Nurs Res. 1997;6:363-74.

5. Martinez LR, Han G, Chacko M, et al. Antimicrobial and healing efficacy of sustained release nitric oxide nanoparticles against Staphylococcus aureus skin infection. *J Invest Dermatol.* 2009;129(10):2463–2469.

6. Singh R, Ray P, Das A, Sharma M. Role of persisters and small-colony variants in antibiotic resistance of planktonic and biofilm-associated *Staphylococcus aureus*: an in vitro study. *J Med Microbiology.* 2009;58(8):1067-1073.

7. Chule K, Ling YC, et al. Preparation and characterization of ZnO nanoparticles coated paper and its antibacterial activity study. Green Chemistry. 2006;8(12):1034-1041.

8. Chiang WC, Schroll C, et al. Silver-palladium surfaces inhibit biofilm formation, *Appl. Environ. Microbiol.* 2009; 75(6):1674.

9. Kutsal Devrim Secinti, Hakan Ozalp, et al. Nanoparticle silver ion coatings inhibit biofilm formation on titanium implants. *Journal of Clinical Neuroscience.* 2011; 18: 391-395.

10. Yuan Liao, Yaqin Wang, et al. Antibacterial surfaces through dopamine functionalization and silver nanoparticle immobilization. *Materials Chemistry and Physics.* 2010; 121: 534-540.

11. Robert JC, Timothy JN Smith, et al. Antibacterial activity of multilayer silver-copper surface films on catheter material. *Canadian Journal of Microbiology.* 1993; 39(9): 895-899

12. Mulligan AM, Wilson M, Knowels JC. The effect of increasing copper content in phosphate-based glasses on biofilms of Streptococcus sanguis. *Biomaterials.* 2003; 24: 1797-1807.

13. Brewer GJ. Copper toxicity in the general population. *Clin Neurophysiol.* 2010; 121(4):459-460.

14. Foldbjerg R, Autrup H, et al. PVP-coated silver nanoparticles and silver ions induce reactive oxygen species, apoptosis and necrosis in THP-1 mono-cytes. *Toxicol. Lett.* 2009; 190:156-162.

15. Xia T, Nel AE, et al. Comparison of the Mechanism of Toxicity of Zinc Oxide and Cerium Oxide Nanoparticles Based on Dissolution and Oxidative Stress Properties. ACS Nano. 2008;2(10):2121-2134.

16. Kryukov GV, Castellano S, Novoselov SV, et al. Characterization of mammalian selenoproteomes. *Science* 2003; 300: 1439 – 43.

(210a) Characterization of Coherent Feedforward Motifs in Mammalian Cells Using Synthetic Gene Circuits

Richard Moore[1], Li Yi[1] and Leonidas Bleris[1,2], (1)Bioengineering, The University of Texas at Dallas, Richardson, TX, (2)Electrical Engineeering, The University of Texas at Dallas, Richardson, TX

Characterization of Coherent Feedforward Motifs in Mammalian Cells using Synthetic Gene Circuits

It is by now appreciated that the behavior of large biochemical networks can often be understood through the analysis of relatively small network motifs. Synthetic circuit engineering gives us the ability to use only a handful of components to build a range of small-scale networks. We report the construction and characterization of three (well-studied in bacteria) feedforward motifs, which we introduce in mammalian cells via transient transfections. Each motif consists of three nodes which code for transcription factors that interact to form the following specific architectures: a type 1 feedforward loop with OR decision at the final node, a type 1 feedforward loop with AND decision, and a cascade style architecture. We selected components that are known to have minimum interference with the host, and designed all necessary control experiments. Additionally, the edges to the second and final node are controlled by ligands, giving us the flexibility to study these networks under a range of different conditions.

We monitor each node via the use of the three different fluorescent reporters, which have negligible crosstalk (microscopy and flow cytometer measurements). Using control experiments, we have established that the AND and OR decisions indeed operate as expected at the final node for the respective architectures. In addition, we have established that the cascade shows a clear delay from the initial node to the secondary node and from the secondary node to the final node, under saturating levels of ligands. We will present time-lapse microscopy and flow cytometry data, for a range of induction levels and perturbations, and will discuss the observed properties of these feedforward motifs. Furthermore, we use population and single-cell data to build quantitative models of the architectures' characteristics. To conclude, the topologies of the circuits that we study are found in many types of natural mammalian networks, including neural and transcriptional networks. We will discuss the relevance to endogenous networks and the implications to understanding and mapping biological pathways.

(210b) High Throughput Monitoring of Pathway Activation Upon Ectopic Expression of Nanog in Human Mesenchymal Stem Cells Using Lentiviral Arrays

Janhavi Moharil[1], Panagiotis Mistriotis[1], Hui You[1], Pedro Lei[1], Jun Tian[1] and Stelios T. Andreadis[1,2], (1)Chemical and Biological Engineering, University at Buffalo, State University of New York, Amherst, NY, (2)Center of Excellence in Bioinformatics and Life Sciences, Buffalo, NY

Nanog has been shown to regulate the proliferation capacity of Mesenchymal Stem Cells (MSCs) and alter their differentiation potential as assessed by overexpression and knock-down studies. Overexpression of Nanog in MSCs is known to reduce their adipogenic differentiation potential while inducing osteogenesis. Our laboratory has shown that Nanog overexpression in MSCs enhances their myogenic differentiation potential as evidenced by improved contractility. However the molecular mechanisms by which Nanog plays a specific role in differentiation are not known. In this study we present a novel tool for high throughput screening of different pathways in Nanog expressing MSCs. We employed a novel dual-promoter lentiviral vector (LVDP) carrying human PGK promoter driving the expression of DsRed serving as an internal control and the promoter of interest driving the expression of ZsGreen to monitor gene expression upon induction of Nanog expression in human Hair Follicle derived MSCs (hHF-MSCs) and Bone Marrow derived MSCs (hBM-MSCs). We employed a Tet-On system for induced expression of Nanog in hHF-MSCs and hBM-MSCs in the presence of doxycycline. The MSCs were transduced with a lentiviral library of about 60 vectors containing promoters and transcriptional response elements (RE) targeting about 25 different pathways and lineage specific markers. The cells were subsequently induced to myogenic differentiation and the promoter/RE activity was monitored over time with automated fluorescence microscope. Fluorescence intensity was quantified by Cell Profiler and the green fluorescence intensity was normalized by red fluorescence intensity. First, we used the lentiviral library to monitor gene expression dynamics during myogenic differentiation of hHF-MSCs and hBM-MSCs over a period of 7 days. Smooth muscle cell markers such as α-SMA, Myocardin, RE-CArG, and components of TGF-β1 signaling pathway such as RE-SMAD2/3 and RE-SMAD7 showed significant upregulation. In the presence of TGF-β1 receptor inhibitor SB431542, RE-SMAD2/3 and RE-SMAD7 did not respond, indicating that the pathway was shutdown as expected. We then employed the lentiviral array to identify which pathways were activated upon ectopic expression of Nanog. Overexpression of Nanog alone induced a marked response in RE-CArG within 3 days showing an enhanced potential for myogenic differentiation (**Figure 1**). Treatment of $Nanog^+$ cells with TGF-β1 and Heparin further increased the activity of RE-CArG, suggesting that effects of Nanog and TGF-β1 in promoting MSC myogenesis were additive. RE-ERK representing the MAPK signaling pathway showed an increased activity in $Nanog^+$ cells. However treatment of $Nanog^+$ cells with TGF-β1 blocked its response. RE-STAT3 (JAK/STAT signaling pathway), RE-KLF4, RE-GATA and RE-GATA2 also showed significant activation in $Nanog^+$ cells but treatment with TGF-β1 did not increase their response any further. We thus demonstrate the application of lentiviral arrays in monitoring gene expression during myogenic differentiation of hHF-MSCs and hBM-MSCs. We present a novel screening strategy to identify various pathways that are activated upon ectopic expression of Nanog in MSCs. Currently we are applying this approach to study the role of Nanog in osteogenic, chondrogenic and adipogenic differentiation.

Figure 1:.Heatmap representation of high throughput array. The figure shows promoter / response element activity in HF-MSCs and Nanog+ HF-MSCs in growth media and myogenic differentiation media after 3 days of treatment. The response in each condition was normalized to the response of HF-MSCs in growth media (control). The dendrogram was computed by hierarchical clustering.

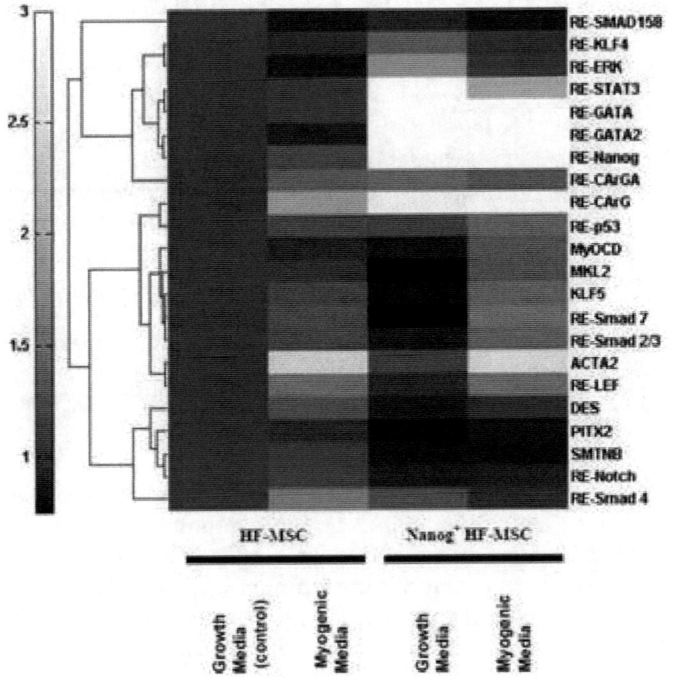

(210c) Integrated Single-Cell Analysis of Heterogeneous Secretory Profiles Exhibited by Human Primary Colorectal Tumor Cells

Viktor Adalsteinsson[1,2,3], Naren Tallapragada[1,4], Narmin Tahirova[1], Xiaosai Yao[1,5], Alessandro Angelini[1], K. Dane Wittrup[1,3,5] and J. Christopher Love[1,2,3], (1)The David H. Koch Institute for Integrative Cancer Research, Massachusetts Institute of Technology, Cambridge, MA, (2)The Broad Institute of Harvard and MIT, Cambridge, MA, (3)Department of Chemical Engineering, Massachusetts Institute of Technology, Cambridge, MA, (4)Department of Electrical Engineering and Computer Science, Massachusetts Institute of Technology, Cambridge, MA, (5)Department of Biological Engineering, Massachusetts Institute of Technology, Cambridge, MA

There is increasing evidence for genetic heterogeneity among individual tumor cells. In colon cancer, multilineage differentiation and epigenetic regulation have also been shown to induce transcriptional heterogeneity, which may affect other complex functional behaviors such as secretion of growth factors or chemokines. Since secreted factors produced by tumor and stromal cells can affect progression of disease, it is important to understand the functional heterogeneity that exists within the tumor microenvironment in order to begin mapping signaling networks. Bulk methods such as ELISAs or protein microarrays lack the ability to resolve functional heterogeneity among single cells. In this study, we employ quantitative microengraving, in combination with on-chip cytometry, to probe the secretomes of over 10,000 human primary colorectal tumor cells in parallel using arrays of subnanoliter wells (nanowells). Starting with a panel of 21 cytokines, chemokines, and growth factors relevant to cancer biology, we examined protein production from six human primary tumors and identified common responses among the samples, including chemokines such as IL-8. We show that individual cells from the tumors exhibit discrete secretory profiles, suggesting that functional heterogeneity is common among cells from a single tumor. We also find evidence for intercellular bypass signaling via production of redundant chemokines that agonize the same receptor.

(210d) Chance and Circumstance Govern Macrophage Functional Diversity

Yishan Chuang, Chemical and Biologial Engineering, Northwestern University, Evanston, IL and Joshua N. Leonard, Chemical and Biological Engineering, Chemistry of Life Processes Institute, R.H. Lurie Comprehensive Cancer Center, Northwestern University, Evanston, IL

Macrophages play a critical role in maintaining the balance between homeostasis and protective inflammation by adopting either immunostimulatory (M1) or immunosuppressive (M2) phenotypes. Tumors and pathogens manipulate this balance by inducing the production of immunosuppressive stimuli, which shifts macrophages into a M2 phenotype, supporting tumor growth and pathogen survival. In addition, M2 macrophages have been reported to convert to an M1 phenotype given certain stimuli, which is termed "plasticity". Despite the centrality of these phenomena to diverse disease processes, the mechanisms regulating macrophage "decision-making" vis-à-vis polarization remain poorly understood. Here, we describe several novel systems-level insights into how such dysfunctional immune network states are established, maintained, and potentially may be disrupted.

Since most macrophage polarization studies to date involve applying coherent stimuli (either pro-M1 or pro-M2 stimuli alone), and these stimuli rarely exist independently *in vivo*, we first investigated how macrophages integrate contradictory signals to "calculate" a response. Macrophages were exposed to combinatorial variations in dose of IL-10 and IL-12 (pro-M2 and pro-M1 stimuli, respectively), and macrophages were "activated" using bacterial endotoxin. Surprisingly, the presence of IL-10 blocked previously-described IL-12-mediated conversion of M2 cells to an M1 phenotype. Moreover, M2-type responses increased with IL-10 dose and were largely independent of IL-12 co-treatment. We also investigated a competition between IL-10 and the canonical pro-M1 stimulus, IFNγ. Interestingly, IFNγ specifically suppressed the production of additional IL-10 by M2 cells, potentially disrupting a key positive feedback loop involved with maintaining the M2 state. This suggests that IFNγ may enable the restoration of an immune network state that promotes M1 polarization and immunological clearance or control of cancer.

We also investigated whether incoherent stimuli induce heterogeneous responses among different macrophages within a single population. Using flow cytometric profiling, we observed that under some conditions, distinct M1 and M2 cells were induced within a single population. The probability of polarization towards an M2 state increased with IL-10 dose and was independent of IL-12 co-treatment. Other cells remained non-responsive to activation, and this probability of activation was unaffected by cytokine treatment. These data represent the first evidence to date that macrophage polarization may be a stochastic process and suggest that separate regulatory mechanisms govern activation and polarization. Subsequent investigations focus on determining whether plasticity is also heterogeneous, and if so, how this is dependent upon a cell's prior polarization history. This system-level analysis of macrophage function provides new insights into innate immune function and will help to formulate therapeutic strategies for treating the many chronic diseases characterized by local immune dysfunction.

(210e) Multiscale Prediction of Patient-Specific Platelet Function Under Flow

Matthew H. Flamm[1], Thomas Colace[1], Manash S. Chatterjee[1], Hiuyan Jing[2], Songtao Zhou[2], Daniel Jaeger[2], Lawrence F. Brass[3], Talid R. Sinno[1] and **Scott L. Diamond**[1], (1)Chemical and Biomolecular Engineering, University of Pennsylvania, Philadelphia, PA, (2)University of Pennsylvania, Philadelphia, PA, (3)School of Medicine, University of Pennsylvania, Philadelphia, PA

During thrombotic or hemostatic episodes, platelets bind collagen and release ADP and thromboxane A2, recruiting additional platelets to a growing deposit that distorts the flow field. Prediction of clotting function under hemodynamic conditions for an individual's platelet phenotype remains a challenge. A platelet signaling phenotype was obtained for 3 healthy donors using Pairwise Agonist Scanning (PAS), where calcium dyeloaded platelets were exposed to pairwise combinations of ADP, U46619, and convulxin to activate P2Y1/P2Y12, TP, and GPVI receptors, respectively, in the presence or absence of the IP receptor agonist, iloprost. A neural network model was trained on each donor's PAS experiment and then was embedded into a multiscale Monte Carlo simulation of donor-specific platelet deposition under flow. The simulations were directly compared to microfluidic experiments of whole blood flowing over collagen at 200 and 1000 s-1 wall shear rate. The simulations predicted the ranked order of drug sensitivity for indomethacin, aspirin, MRS-2179 (P2Y1 inhibitor), and iloprost. Consistent with measurement and simulation, one donor displayed larger clots, while another donor presented indomethacin-resistance (revealing a novel heterozygote TP-V241G mutation). *In silico* representations of an individual's platelet phenotype allowed prediction of blood function under flow, essential to identifying patient-specific cardiovascular risks, drug responses, and novel genotypes.

Figure 1

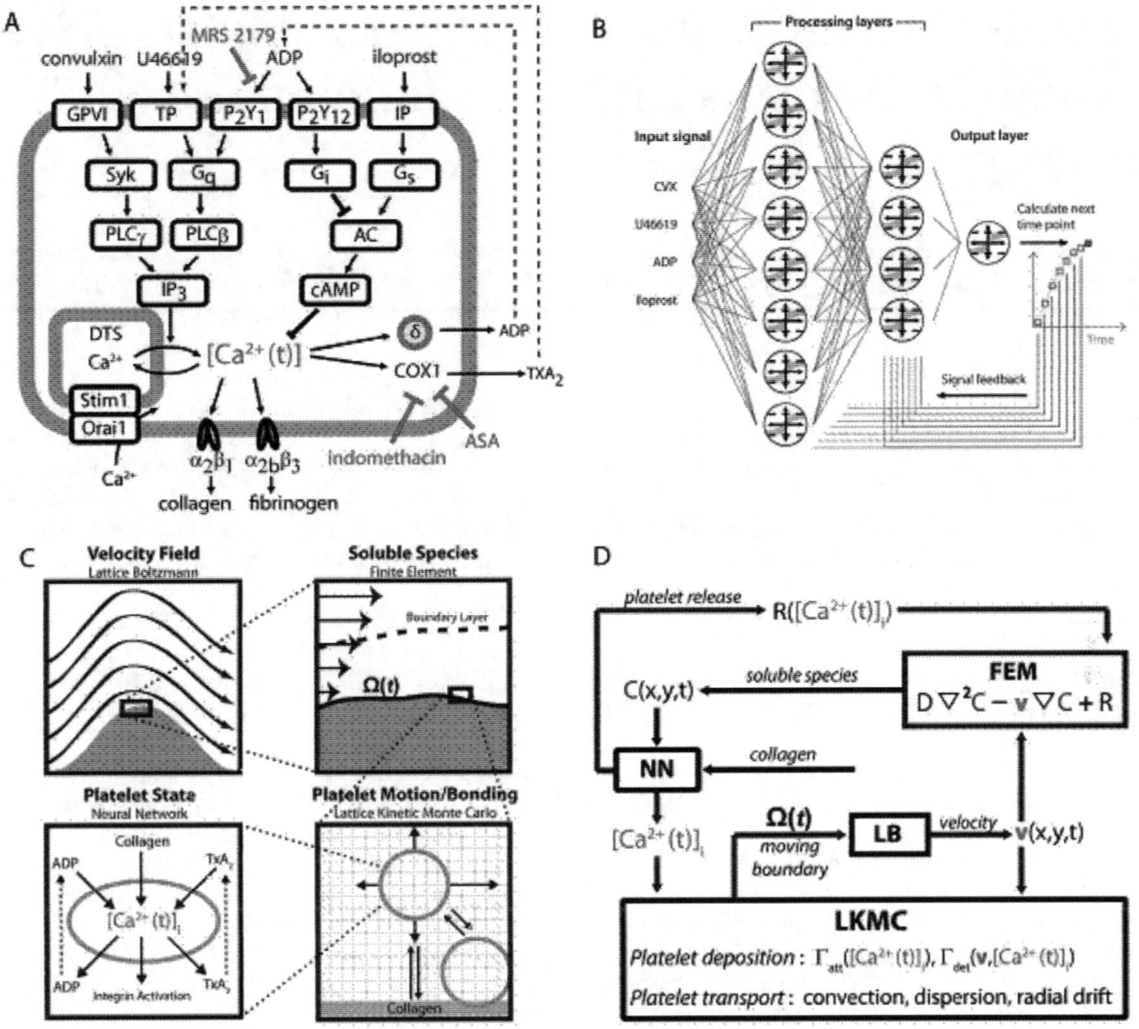

Fig. 1. Multiscale model of combinatorial platelet activation and thrombus formation under flow. Platelet agonists (*blue*) used individually or in pairs to activate GPVI or G-protein coupled receptors (thromboxane receptor, TP; purinergic receptors P2Y1 and P2Y12; and the prostacyclin receptor, IP) result in modulation of intracellular calcium (*green*) from intracellular stores distal of phospholipase C (PLC) activation or from store operated calcium entry via Stim1-Orai1 activation. Inhibitors (*red*) such as acetylsalicylic acid (ASA) or indomethacin inhibit cyclooxygenase-1 (COX-1). Autocrine pathways include release of TXA2 and ADP (**A**). A 2-layer, 8-node/4-node neural network (NN) with feedback is trained with 74 measured calcium traces to predict [Ca]i for each patient-specific platelet Pairwise Agonist Scan (**B**). The multiscale simulation of platelet deposition under flow requires simultaneous solution of the instantaneous velocity field over a complex and evolving platelet boundary Ω(t) by Lattice Boltzmann (LB), concentration fields of ADP and TXA2 by finite element method (FEM), individual intracellular platelet state ([Ca]i) and release reactions (R) for ADP and TXA2 by neural network, and all platelet positions and adhesion/detachment by lattice Kinetic Monte Carlo (LKMC) (**C,D**).

(210f) Prediction and Validation of Microbiota-Derived Tryptophan Metabolites with Anti-Inflammatory Properties

Kyungoh Choi[1], Gautham V. Sridharan[2], Robert Alaniz[3], Kyongbum Lee[2] and Arul Jayaraman[1], (1)Department of Chemical Engineering, Texas A&M University, College Station, TX, (2)Chemical and Biological Engineering, Tufts University, Medford, MA, (3)Department of Microbial and Molecular Pathogenesis, Texas A&M Health Science Center, College Station, TX

The human gastrointestinal (GI) tract is colonized by ~10^{14} bacteria belonging to ~ 500 species that are collectively termed the microbiota. Several clinical and *in vivo* studies have shown that alterations in the microbiota composition and/or function (i.e., its dysbiosis) are correlated to pathologies such as inflammatory bowel disease (IBD) and colitis. The importance of microbiota metabolites is also evident from a metabolomic study of patients with Crohn's disease where disease severity was correlated with alterations in the levels of tryptophan in feces. Recently, work from our lab has shown that indole, a metabolite derived by the microbiota from tryptophan, attenuates indicators of inflammation intestinal epithelial cell. These studies highlight the potentially important role for tryptophan metabolites in the regulation of inflammation in the GI tract. However, identification of metabolites that exist in the GI tract is very difficult because many of the gut microbiota species cannot be cultured *in vitro*. As metabolic pathways are highly conserved across different organisms and many of the microbiota metabolites can also be synthesized by host cells, the panel of metabolites that can be exclusively synthesized by the microbiota is largely unknown.

In this study, we developed a computational approach for predicting possible tryptophan-derived metabolites that can be uniquely ascribed to microbiota using probabilistic pathway analysis. The probabilistic pathway construction algorithm explores the space of possible biochemical transformations in the KEGG database from a starting metabolite (tryptophan) and predicts all possible product metabolites that can be formed from reactions catalyzed by enzymes that are expressed exclusively in bacteria but not in host cells (e.g. human or mouse). Each of these product metabolites, in turn, are used as substrates for the next round of possible biotransformations, and the algorithm continues until a product metabolite that cannot be uniquely ascribed to bacteria is reached. Starting from tryptophan and its immediate metabolites, the algorithm predicted ~10 possible metabolites that can be uniquely ascribed to microbiota reactions. A targeted metabolomics approach was then taken to identify and quantify four of the predicted metabolites – indole-3-acetate, indole-3-pyruvate, indole-3-acetamide, and tryptamine - that were present in murine intestinal tissue and fecal samples using an LC/MS-MS system operating in multiple reaction monitoring (MRM) mode. Preliminary data suggest that the identified metabolites attenuate indicators of inflammation in mucosal cells; thereby, demonstrating the potential and utility of our integrated bioinformatics and metabolomics approach. The identification of novel anti-inflammatory metabolites is expected to lead to impact the development of therapeutic molecules against inflammatory disorders such as IBD and colitis.

(210g) Understanding Nitrogen Metabolism in the Biofuel Crop Poplar by Isotope-Assisted Metabolic Flux Analysis

Shilpa Nargund[1], Ashish Misra[1], Xiaofeng Zhang[1], Gary D. Coleman[2] and Ganesh Sriram[1], (1)Chemical and Biomolecular Engineering, University of Maryland, College Park, MD, (2)Plant Science and Landscape Architecture, University of Maryland, College Park, College Park, MD

Nitrogen crucially determines the growth and productivity of photosynthetic organisms. However, it is often supplied by means of artificial fertilizers, whose production is energetically intensive and significantly relies on non-renewable fossil fuels. Therefore, the investigation of nitrogen use efficiency (NUE), particularly in plants that are potential biofuel crops, will have wide-ranging consequences for biofuel economy and sustainability. Poplar, a biofuel crop, uses an array of sophisticated metabolic and regulatory mechanisms to deposit, remobilize and thereby conserve nitrogen. The transport of nitrogen from senescing leaves to the bark storage protein (BSP) during winter is one such mechanism. This recycling of nitrogen is closely related to the status of carbon availability in plants as evidenced by previous studies which show that photoperiod perception, growth, presence of glutamine or calcium and protein phosphorylation affect accumulation of BSP (Zhu. B and Coleman. G.D., Plant Physiology 126:342-351, 2001). The involvement of multiple factors in regulation of BSP production suggests that significant metabolic rewiring occurs during nitrogen recycling.

To study this phenomenon we employed metabolic engineering tools such as steady state isotope-assisted metabolic flux analysis (MFA) to quantify metabolic fluxes in cells of the tree poplar (*Populus trichocarpa*). Metabolic fluxes are the rates of carbon flow through metabolic pathways (measured in mol $time^{-1}$ g $biomass^{-1}$) and are important indicators of cellular functions (Stephanopoulos, G. & Stafford, D.E., Chemical Engineering Science 57: 2595-2602, 2002). MFA involves conducting isotope labeling experiments (ILEs) wherein isotopically labeled (e.g. ^{13}C, ^{15}N, ^{17}O) substrates are fed to the plant cell culture or tissue and the labeling patterns of biomass components such as proteins, lipids, sugars and nucleotides, are measured using nuclear magnetic resonance (NMR) or mass spectrometry (MS). Fluxes are estimated by fitting the labeling patterns to a mathematical model of the metabolic network.

Poplar suspension cells, subjected to environmental or genetic perturbations that affect BSP metabolism, will be grown under 100% 1-^{13}C and 30% U-^{13}C glucose in separate ILEs. Comparative metabolic flux maps for poplar cells under the various treatments will be constructed to evaluate the changes in metabolism during BSP accumulation and N recycling. The differences in fluxes between conditions and their implications on cell physiology will be discussed. The results of our MFA experiments are anticipated to reveal the metabolic rewiring that occurs in poplar cells when they deposit and remobilize nitrogen, and eventually provide insights toward engineering photosynthetic organisms with high NUE.

(210h) Tandem Mass Spectrometry: The Next Frontier in 13C-Metabolic Flux Analysis

Maciek R. Antoniewicz, Department of Chemical and Biomolecular Engineering, University of Delaware, Newark, DE and Jungik Choi, Chemical Engineering, Univerisity of Delaware, Newark, DE

Tandem mass spectrometry is a new measurement technique for quantifying isotopic labeling for ^{13}C-metabolic flux analysis (^{13}C-MFA). Recently, we developed a general mathematical framework based on isotopomer simulations for incorporating tandem MS data into ^{13}C-MFA (Choi et al., Metab Eng 2011). Here, we present further advancements we have made towards establishing next-generation tandem MS based flux analysis, specifically: 1) measuring complete positional isotopomer distribution of aspartate by tandem MS; and 2) developing an elementary metabolite units (EMU) framework for modeling tandem mass isotopomers.

First, we developed a simple and accurate method for determining the complete positional isotopomer distribution of aspartate carbon atoms by gas chromatography tandem mass spectrometry (GC-MS/MS). To achieve this, we screened tandem MS spectra of the TBDMS-derivative of aspartate for daughter fragments with the necessary carbon atom fragmentations to fully determine all sixteen isotopomers of aspartate. We selected five fragments that provided a redundant set of labeling measurements to quantify all isotopomers by least-squares regression. We demonstrated the validity of our approach using ^{13}C-labeled aspartate standards and natural aspartate.

Next, we extended our modeling framework for tandem MS simulations. Here, we present for the first time our new framework based on EMUs, which are computationally more efficient to evaluate compared to previous methods based on isotopomers/cumomers. The new mathematical framework, which we called EMU^2, is based on two-dimensional representation of isotopic-labeling. Using this new formulation we take full advantage of the additional information that is obtained from tandem MS measurements for ^{13}C-MFA.

(230a) Engineering Single Wall Carbon Nanotubes for Sub-Cellular Delivery

Brian D. Holt[1], Patrick D. Boyer[2], Kris Noel Dahl[1,2] and **Mohammad F. Islam**[3], (1)Biomedical Engineering, Carnegie Mellon University, Pittsburgh, PA, (2)Chemical Engineering, Carnegie Mellon University, Pittsburgh, PA, (3)Materials Science & Engineering, Carnegie Mellon University, Pittsburgh, PA

As part of nanomaterial applications in biology and medicine, we are interested in developing single wall carbon nanotubes (SWCNTs) as intracellular biomaterials. Our research in this area has been directed to characterizing, developing, and utilizing SWCNT functionalization to control SWCNT sub-cellular localization and probe fundamental cellular processes utilizing SWCNTs. To produce biologically relevant solutions of individualized SWCNTs and overcome aggregation of SWCNTs, we use a surfactant dispersant. Specifically, we are focused on developing bio-inert, biological, and bioactive SWCNTs dispersions to understand and control how the SWCNT dispersions enter cells, where they go and what they do once inside cells, and do they leave cells.

The first step in targeted cell delivery is to determine how SWCNTs get into cells. To determine if SWCNTs dispersed in Pluronic F127 (PF127) could physically penetrate a cell membrane, we performed complementary studies with *in vitro* synthetic membranes, plasma membrane vesicles and live cell experiments. Our results demonstrated that, while P-F127 dispersed SWCNTs are membrane active and associate with cell membranes, they lack sufficient energy to physically penetrate a cell membrane bilayer. Further, our experiments suggested that SWCNTs are directly present in endosomes. Therefore, we experimentally validated that SWCNTs enter cells *via* endocytosis and not membrane penetration.

Once inside the cell, we were interested in the sub-cellular localization and cellular effects of SWCNTs. We showed that highly purified, length-selected, and SWCNTs well-dispersed with PF127 (bio-inert) are not acutely cytotoxic but reduce proliferation in a dose-dependent manner. Due to the similar mechanical and anisotropic properties of SWCNTs and F-actin, we investigated the interaction and effects of SWCNTs on F-actin. These SWCNTs reorganized F-actin structures, leading to more F-actin intensity in the apical cellular regions. Fluorescence lifetime imaging microscopy further suggested that SWCNTs interact with F-actin structures both in the cells and in purified actin filaments reconstituted *in vitro*. Functionally, SWCNTs reduce forces that can be generated by NIH-3T3.

Endocytosed SWCNTs coated with PF127 were able to enter cellular compartments and interact with actin, likely from their membrane activity which could destabilize endosomes. However, dispersion with bovine serum albumin (BSA) caused markedly different cellular effects. SWCNTs-BSA were not acutely cytotoxic and did not significantly reduce proliferation. Confocal Raman imaging co-registered to phase contrast showed different patterns of sub-cellular localization for these dispersions within perinuclear region of the cell. We also observed a significant increase in cellular uptake of SWCNTs-BSA into cells with 10's of millions of SWCNTs entering human mesenchymal stem (hMSC) cells and HeLa and NIH-3T3 cells. Surprisingly, steady-state uptake per cell occurs in less than 1 minute, and there is a concentration threshold between 1 and 30 mg/mL as cell uptake machinery is saturated. Confocal Raman mapping of SWCNT radial breathing modes, fluorescence, and G-band demonstrated that SWCNTs primarily remain individually dispersed.

Finally, we determined that cells could recover from SWCNT exposure. Most cell lines exposed to SWCNTs-BSA were able to expel SWCNTs and recover cell phenotype, and after ~80 hr essentially no SWCNTs remained in cells. Conversely, cells exposed to SWCNTs-PF127 did not recover, maintaining substantial SWCNTs-PF127 throughout the cells, even after 80 hr. This differential ability to recover suggests different subcellular compartmentalization of SWCNTs as well as the possibility for cellular delivery of SWCNTs within the body.

In conclusion, we have dispersed SWCNTs – an important nanostructured material for biological applications – with different classes of biological dispersing agents to control uptake, sub-cellular processing, and recovery. Therefore, from these studies, we have shown that we can engineer and control SWCNT-cell interactions.

References

1. B. D. Holt, K. N. Dahl, and M. F. Islam, "Cells Uptake and Recover from Protein Stabilized Single Wall Carbon Nanotubes with Two Distinct Rates", *ACS Nano* **6**, 3481 (2012).
2. P. N. Yaron, B. D. Holt, P. A. Short, M. Lösche, M. F. Islam, and K. N. Dahl, "Single wall carbon nanotubes enter cells by endocytosis and not membrane penetration", *Journal of Nanobiotechnology* **9**, 45 (2011).
3. B. D. Holt, K. N. Dahl, and M. F. Islam, "Quantification of Uptake and Localization of Bovine Serum Albumin-Stabilized Single-Wall Carbon Nanotubes in Different Human Cell Types", *Small* **7**, 2348 (2011).
4. B. D. Holt, P. A. Short, A. D. Rape, Y. Wang, M. F. Islam, and K. N. Dahl, "Carbon Nanotube Reorganize Actin Structures in Cells and *ex vivo*", *ACS Nano* **4**, 4872 (2010).

(230b) Nanocapsules for Protein Delivery

Yunfeng Lu, Chemical and Biomolecular Engineering, University of California at Los Angeles, Los angeles, CA

An average cell contains thousands of proteins participating in its normal cellular functions. Most diseases, however, are somehow related to the malfunction of specific proteins. Protein therapy, endowed with more defined action mechanism than small molecule drugs and safer administration methods than gene therapy, is considered to be a treatment modality of choice for many disorders today and in the future. Advances in biotechnology have accelerated the economical, large-scale production of proteins, vaccines, and hormones, making them readily available for therapeutic applications in medical practices and clinical studies. Up to now, over 140 FDA-approved protein drugs have been placed on the market, and more are coming. Despite its fast growth in the recent years, protein therapy still faces critical challenges, such as poor *in-vitro* and *in-vivo* stability, fast degradation and clearance, and the incapability to delivery proteins across biological barriers. Developing efficient protein delivery strategy is still in high demand. Various factors affect the efficiency of protein delivery agents, including surface properties, size, and shape. To develop protein delivery agents for various therapeutic applications, we designed a general strategy based on protein nano-encapsulation, a novel protein-polymer conjugation technique achieved by in-situ polymerization on protein surface. This approach yielded protein nanocapsules with retained activity, enhanced stability and the ability to transport through biological barriers (such as cell membranes). This strategy is applicable over a variety of proteins, for a wide spectrum of therapeutic applications. Our in-vitro and in-vivo experiments demonstrated the potential of the nano-encapsulation strategy for the treatment of gout, phenylketonuria, alcohol intoxication, and other therapeutic applications.

(230c) Biomimetic Long-Circulating Nanoparticles for Combinatorial Drug Delivery

Liangfang Zhang, Nanoengineering Department, University of California, San Diego, La Jolla, CA

Efforts to extend nanoparticle residence time in vivo have inspired many strategies in particle surface modifications to bypass macrophage uptake and systemic clearance. Herein I report a top-down biomimetic approach in particle functionalization by coating biodegradable polymeric nanoparticles with natural red blood cell (RBC) membranes including both membrane lipids and associated membrane proteins for long-circulating cargo delivery. This approach aims to camouflage the nanoparticle surface with the erythrocyte exterior for long circulation while retaining the applicability of the polymeric core. In vivo results revealed superior pharmacokinetics and biodistribution by the RBC-mimicking nanoparticles as compared to control particles coated with the state-of-the-art synthetic stealth materials. I will also report several new strategies for nanoparticle-based combination chemotherapy by concurrently incorporating two different types of drugs into a single polymeric nanoparticle with ratiometric control over the loading of the two drugs. The cytotoxicity of these dual-drug carrying nanoparticles was compared with their cocktail mixtures of single-drug loaded nanoparticles and showed superior therapeutic effect.

(230d) Corking Carbon Nanotube Cups with Gold Nanoparticles for Drug Delivery Applications

Alexander Star, Department of Chemistry, University of Pittsburgh, Pittsburgh, PA

In this talk, I will describe an effective methodology of obtaining nitrogen-doped carbon nanotube cups (NCNCs), which have a hollow multiwall graphitic structure with one end sealed and the other open. We have previously shown that nitrogen doping is essential to determine the unique morphology and properties of NCNCs.[1,2] Other studies have demonstrated that nitrogen doped carbon nanotubes have better biocompatibility and mitigated cytotoxicity as compared to undoped carbon nanotubes.[3] These findings are important when considering NCNCs as potential carriers in drug delivery applications. Most recently, we have demonstrated[4] that by applying intense ultrasonication, stacked NCNC fibers, which were prepared by chemical vapor deposition (CVD) technique, can be greatly shortened to afford individual NCNCs with an efficient yield. Microscopic and spectroscopic characterization confirmed the cup-shaped hollow structure of isolated NCNCs and showed their robust chemical reactivity due to amine groups that are preferentially distributed at the open rims of the cups. We were further able to cork isolated nanocups with gold nanoparticles (GNPs) by utilizing the localized amine groups as reactive sites. These nanocup-nanoparticle hybrid structures may find future applications as nanoscale containers and drug delivery carriers.

1. Allen, B. L.; Kichambare, P. D.; Star, A. Synthesis, characterization, and manipulation of nitrogen-doped carbon nanotube cups. *ACS Nano* **2008**, *2*, 1914–1920.
2. Tang, Y.; Allen, B. L.; Kauffman, D. R.; Star, A. Electrocatalytic activity of nitrogen-doped carbon nanotube cups. *J. Am. Chem. Soc.* **2009**, *131*, 13200–13201.
3. Castle, A. B.; Gracia-Espino, E.; Nieto-Delgado, C.; Terrones, H.; Terrones, M.; Hussain, S. Hydroxyl-functionalized and N-doped multiwalled carbon nanotubes decorated with silver nanoparticles preserve cellular function. *ACS Nano* **2011**, *5*, 2458–2466.
4. Zhao, Y.; Tang, Y.; Chen, Y.; Star A. Corking carbon nanotube cups with gold nanoparticles. *Submitted.*

(230e) Low Frictional Mesoporous Titanium Dioxide Film by Geometrical Roughness-Induced Heterogeneous Nanostructure From Titanate

Rong An, Xiaohua Lu, Ximing Wu, Changsong Wang, Chang Liu and Shuangqin Fu, chemical engineering, Nanjing university of technology, Nanjing, China

It is of great importance to construct a low frictional heterogeneous titanium dioxide (TiO_2) surface by physical strategy. Native dense TiO_2 layers prefer to forming on bulk titanium (Ti) substrates. However, the poor frictional TiO_2 on Ti surface is not favorable to fluid flow, resulting in harmful effects including but not limited to thrombus, nonspecific adsorption of proteins and cells, osteoarthritis. Thus surface modification with chemical functional groups was utilized to provide a heterogeneous stable and bacteria-resistant TiO_2 surface. But the inducing poor biocompatibility is of great concern and the functional groups are susceptible to oxidation, hydrolysis or thermal degradation.

An approach to low surface friction by constructing geometrical roughness-induced nano-patterned heterogeneous surfaces could confer adsorbates resistance and maintain the surface biocompatibility. Atomic force microscopy (AFM), with a nanoscale sensitivity, offers an unprecedented opportunity to explore the molecular-level factors that govern friction on various surfaces.

Herein, we reported a simple physical strategy based on titanate by constructing a roughness-induced nano-patterned heterogeneous mesoporous TiO_2 film to induce long-range hydrophobic interactions to further low friction. With AFM measurement, the friction coefficient (0.0026, measured in open system: air environment, at 25°C, relative humidity: ~50%) of mesoporous TiO_2 film is lower than 1/25 of dense one (0.067). Due to the existence of TiO_2(B) (59.7% in weight), mesoporous TiO_2 film keeps the structure thermal stability, further leading to the secondary nano-patterned pores (<15nm). By AFM adhesion analysis, the distribution of adhesion with Si_3N_4 tip on mesoporous TiO_2 surface shows heterogeneous property with a bimodal peak (3nN, 12nN) caused by the presence of the aforesaid secondary nano-patterned pores and TiO_2 particles. While the adhesion distributions on dense TiO_2 surface is single(12nN), showing homogeneous, attributing to TiO_2 particles. In AFM phase results, the secondary nano-patterned pores on mesoporous TiO_2 film induce long-range hydrophobic interactions, leading to a relatively hydrophobic boundary region around each hydrophobic pore patch. The boundary length extension can effectively increase the coverage of the hydrophobic phase (nano-patterned pores) to further reduce friction. The strategy to low mesoporous TiO_2 film friction by geometrical roughness-induced heterogeneous nanostructure from titanate benefits the fluid flow to further block the propensity of various adsorbates to avidly stay on the interfaces and it can maintain TiO_2 native biocompatibility.

(230f) SAM-Modified Microdisc Electrode Arrays (MDEAs) with Functionalized Carbon Nanotubes for Amperometric Biosensors

Anthony Guiseppi-Elie, Center for Bioelectronics, Biosensors and Biochips, Clemson University, Anderson, SC

A novel biosensor for glucose, one that makes use of covalent conjugation of glucose oxidase (GOx) to acid-chopped, chemically modified single-walled carbon nanotubes (SWNT) that were themselves covalently immobilized via alkane thiol SAMs onto gold microdisc electrode arrays (MDEAs) has been constructed and evaluated. Amperometric biotransducers fabricated by chemisorbing self-assembled monolayers (SAM) of cysteamine (CA) and 11-amino-1-undecanethiol (11-AUT) onto 100 mm diameter gold microdisc arrays, resulting in MDEA-Au|CA|SWNT-COOH and MDEA-Au|11-AUT|SWNT-COOH electrodes, were characterized electrochemically using cyclic voltammetry and amperometry in the presence of ferrocene. Subsequent covalent conjugation of GOx to free -COOH groups using EDC NHS hetero bifunctional coupling, produced MDEA-Au|CA|SWNT-CONH-GOx and MDEA-Au|11-AUT|SWNT-CONH-GOx electrodes that were similarly characterized in changing concentrations of glucose. The combination of SWNT at the CA-modified MDEA surface compared to the other electrode surfaces gave a well-resolved voltammogram with highest peak currents (I_{pa} = 5.54 x 10^{-4} A, I_{pc} = 2.34 x 10^{-4} A at 100 mV/s). The resulting amperometric glucose biosensor displayed linear responses between 0.5 and 3.0 mM glucose with a response time of 40s and was superior to constructs using 11-AUT.

(249a) Chemical Engineers and Tissue Engineering

W. Mark Saltzman, Biomedical Engineering, Yale University, New Haven, CT

Tissue engineering is a new field of inquiry, defined about 25 years ago, but it is already impacting medical care. Chemical engineers have been central contributors in this effort. Tissue engineering takes methods and materials from chemical engineering and deploys them for new purposes: bioreactors are used for tissue assembly, polymer materials provide scaffolds for tissue growth, and molecular delivery systems control the release of factors essential for cell survival and differentiation. Accordingly, the basic principles for tissue engineering align closely with chemical engineering science. Here, those basic principles are illustrated in examples of tissue engineering in the skin, the cardiovascular system, and the nervous system. These examples will be used to highlight the role of chemical engineers in the design of novel biomaterials, the delivery of agents that control cell function, and the optimization of vascular self-assembly.

Chemical Engineering Approaches to the Design and Fabrication of Extrusion-Based Functionally Graded Scaffolds Targeting Tissue Engineering Applications

Dilhan M. Kalyon

Stevens Institute of Technology

Introduction

The designing and fabrication of bioresorbable polymeric scaffolds for tissue engineering are challenged by the complex elegance of the structural and compositional gradations found in human tissues [1, 2]. For example, it would be desirable for the bone graft substitutes used in the repair of critical sized defects in long bones like femur and tibia to accommodate their changing porosities and moduli along their transverse and axial directions [3,4]. Therefore, mimicking of such complex gradations found in native tissues can require correspondingly complex gradations in tissue engineering scaffolds. Such gradations could be tailored as three-dimensional distributions in composition, structure and properties. Generation of tissue engineering scaffolds with such gradations are constrained by the available conventional scaffolding methodologies. Examples to fabrication of graded structures involve the generation of multilayered scaffolds that could be fabricated via electrospinning of separate meshes and pressing them under hydraulic pressure [5], layer-by-layer casting, freeze-drying, phase separation, and rapid prototyping techniques, including fused deposition modelling, 3D printing, selective laser sintering, and stereolithography [1, 2, 6-12]. However, additional methods that would allow the reproducible and industrially-scalable grading of bone graft substitutes and tissue engineering scaffolds for a wide range of compositions, porosities and mechanical properties are still needed.

In the fabrication of functionally-graded tissue engineering scaffolds extrusion technologies can play significant roles. Various variants of the extrusion processes, including single screw extrusion, twin screw extrusion using co-rotating or counter-rotating twin screw extruders, coupled to various dies and finishing equipment, offer significant advantages since they can be used to carry out multiple elemental steps of polymer processing [13]. These include the conveying of polymeric particles, melting, pressurization of the melt, devolatilization, shaping of the scaffolds. Furthermore, such extrusion equipment can be integrated with gravimetric and loss-in-weight feeders for solids and infusion and other pumps for liquids which allow precise and time-dependent control of the feeding rates of ingredients. Principally three extrusion–based processes were developed to enable the fabrication of functionally-graded scaffolds:

1. Hybrid process consisting of twin screw extrusion and electrospinning [14-17].
2. Twin screw extrusion and spiral winding [18-20].
3. Co-extrusion [21-25].

These three methods [14-25] are shown in Fig. 1 and were applied to the fabrication of scaffolds which target bone tissue engineering applications, and their biocompatibility was demonstrated employing osteoblasts or adipose or bone marrow derived stem cells. However, the three methods can be readily adopted and used for other tissue engineering applications. In the following the basic chemical engineering tools that can be utilized for the development and optimization of the three methods are outlined along with the descriptions of the capabilities of the three methods to fabricate functionally-graded scaffolds. In all cases poly(caprolactone) with or without solvents was employed as the bioresorbable polymeric matrix.

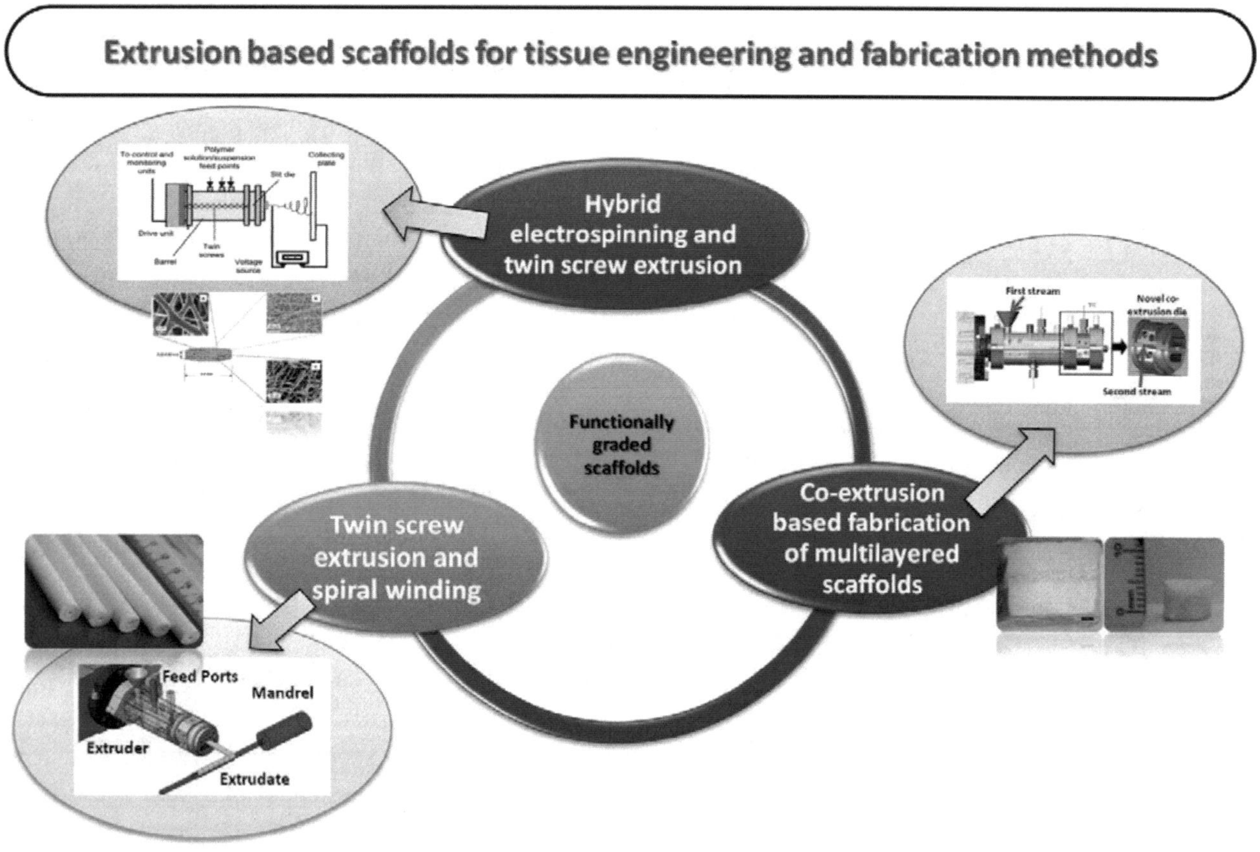

Fig. 1 Three types of functionally-graded scaffolds for tissue engineering applications via co-extrusion, hybrid method of twin screw extrusion and electrospinning and twin screw extrusion and spiral winding.

As summarized in Fig. 1, the twin screw extrusion, TSE, process can be applied concomitantly with electrospinning to generate nanofibrous meshes which can be graded with porosity, pore size and bioactive concentration distributions [14-17]. For example, concentration gradients of two bioactives, ß-gloycerophosphate and insulin could be generated using TSE-electropinning. The regions that are rich in insulin were determined to give rise to chondrocytic differentiation of adipose-derived stem cells, whereas the regions that were rich in ß-gloycerophosphate were determined to furnish greater degree of mineralization [16]. Other examples of extrusion based methods were also developed and applied to generate functionally-graded bone graft substitutes and scaffolds (Fig. 1). In twin screw extrusion and spiral winding the extrudates, the composition and porosity of which can be altered as a function of time, are wound on a simultaneously translating and rotating mandrel to generate radially and axially-graded scaffolds [18-20] and were incorporated with various bioactives with systematic changes in concentration and porosity. Furthermore, axially and radially graded scaffolds of PCL could be fabricated via the co-extrusion method to generate distributions of porosity, pore size and concentration distributions of osteoconductive particles, tricalcium phosphate, TCP, and hydroxyapatite, HA, in both the axial and radial directions (cage/core structures) [21-25] as shown in Fig. 2.

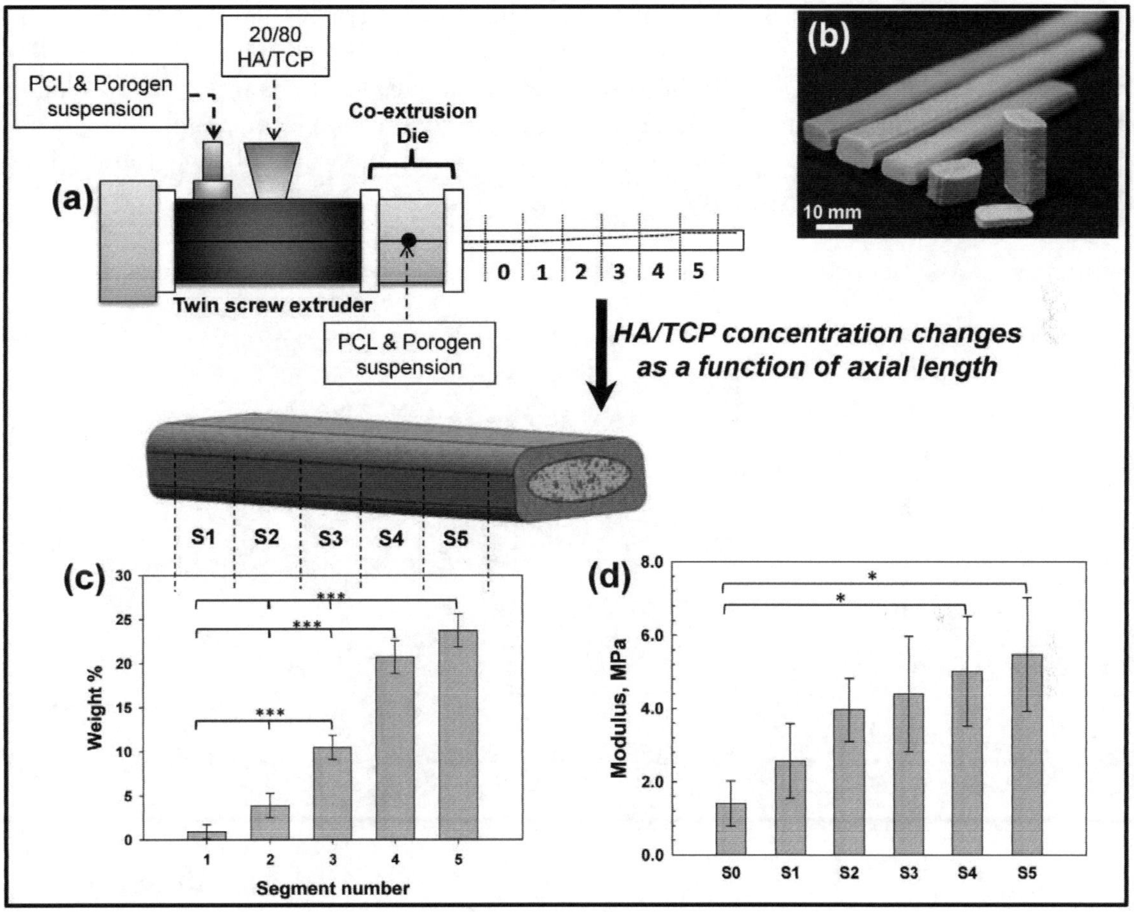

Fig. 2 Functionally-graded scaffolds for bone tissue engineering via co-extrusion with changes in HA/TCP particle concentrations in the axial direction with corresponding changes in the modulus of different segments in the axial direction and porosity gradation in the axial direction [25].

The modulus of the co-extruded scaffolds could be manipulated systematically in both the radial and axial directions. When tissue constructs were generated by seeding bone marrow derived stem cells onto the co-extruded graded scaffolds and samples were harvested as a function of time during cell proliferation and differentiation in media, the gene expression levels and μ-CT analysis revealed the osteogenenic differentiation, bone ECM formation and mineralization as affected by the gradations of the bioactive ingredient, i.e., HA and TCP concentrations [25].

Mathematical modeling of the extrusion flows

In all three methods the common denominator is the use of a twin screw extruder (fully-intermeshing co-rotating twin screw extruder). This is a very versatile extruder which is especially suitable for both distributive and dispersive mixing (thus ideal for mixing of porogens and osteoconductive particles). The basic mixing elements are the kneading disks which can be staggered at various angles in the forward and reverse directions. The following (Fig. 3) shows a schematic of the twin screw extrusion with a slit die in conjunction with two sets of kneading discs staggered at a stagger angle of 90°.

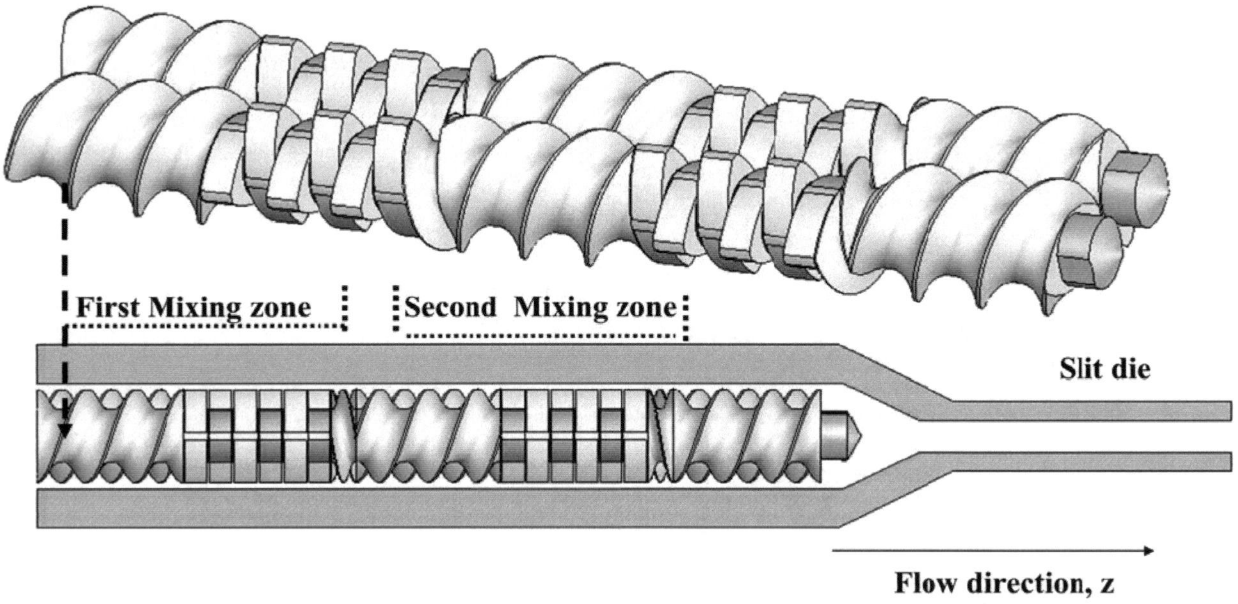

Fig.3 Schematics of the fully-intermeshing co-rotating twin screw extrusion showing a screw configuration involving two sets of kneading disks, regular flighted conveying screw sections and a slit die [34].

The optimum utilization of the extrusion based technologies requires the solution of the conservation equations for mass, momentum and energy in conjunction with appropriate rheological constitutive equations and equation of state. Especially, when salt particles are used as porogens at relatively high concentrations (typically over 60% by volume to generate adequate porosity with interconnected pores) a viscoplastic type constitutive equation can be utilized to represent the flow and deformation behavior of the concentrated suspension to be processed in the extruder (single or twin, co-rotating or counter-rotating, fully-intermeshing or tangential screws). For such filled systems the correct boundary condition is the wall slip condition. Mathematical models based on analytical and numerical solutions of the transport equations for processing of viscoplastic fluids subject to wall slip in various types of extrusion hardware and shaping using various types of extrusion dies are available [26-37]. The characterization of viscoplasticity and wall slip behavior also requires special rheological characterization apparati and methods [38-43].

Figure 4 shows some typical pressure distributions as a function of mass flow rate in the mixing zone of the twin screw extruder obtained via the 3-D Finite Element Method [34]:

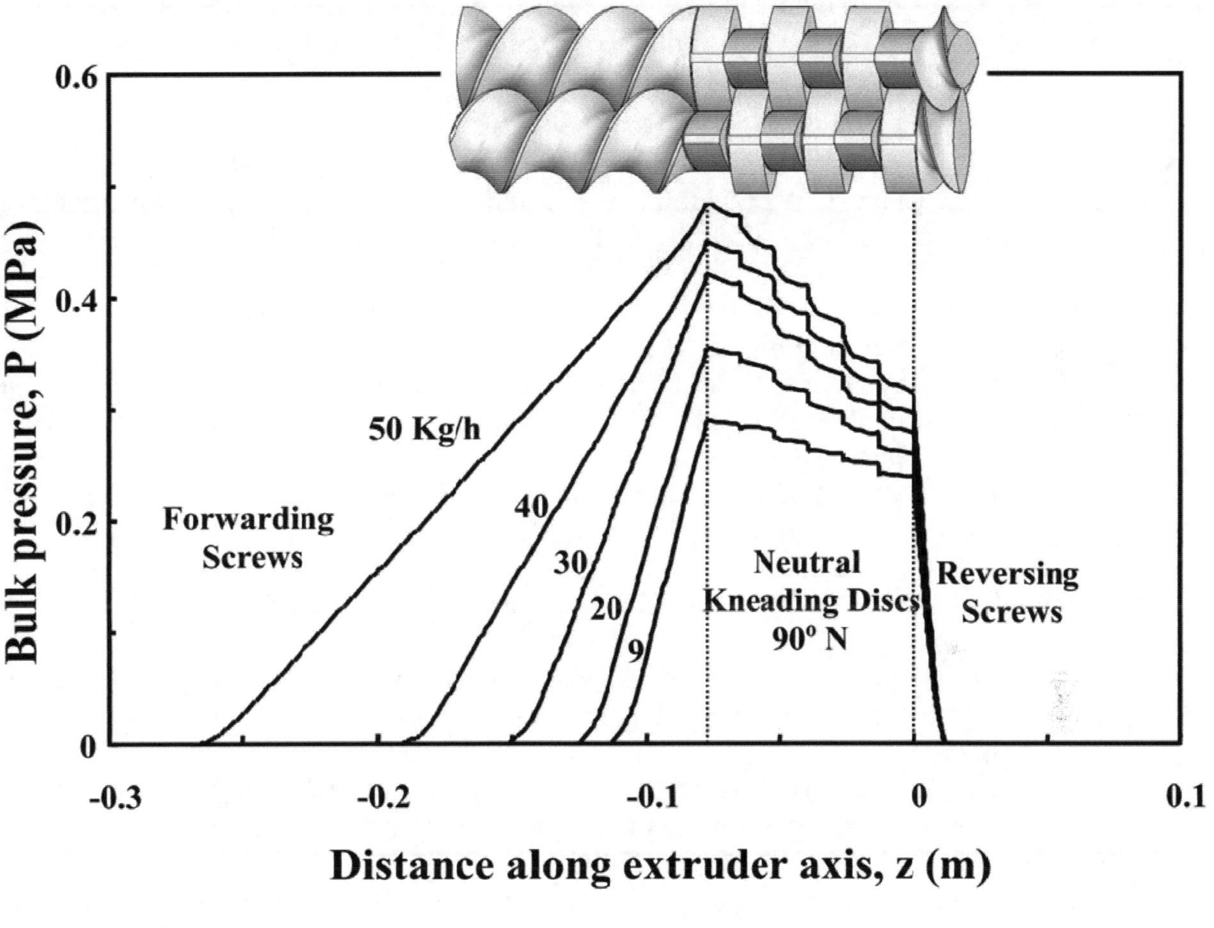

Fig.4 Pressure distributions and degree of fill as a function of flow rate in a mixing section of the fully-intermeshing co-rotating twin screw extruder as a function of mass flow rate [34].

Conclusion

During the last three decades the chemical engineering profession has gained specialized capabilities in rheology, mathematical modeling and simulation on various types of extrusion and fiber spinning methodologies, microstructural analysis and tailoring, and development of mechanical properties to allow the development of extrusion based technologies for the development of bone graft substitutes and tissue engineering scaffolds. The role of chemical engineering is expected to increase with greater utilization of these industrially-relevant methods upon commercial scale implementation to advance patient care by providing better graft substitutes and scaffolds which better mimic the complex elegance of native tissues.

References

1. Leong, K.F., Chua, C.K., Sudarmadji, N., and Yeong, W.Y. Engineering functionally graded tissue engineering scaffolds. J Mech Behav Biomed Mat 1, 140, 2008.
2. Pompe, W., Worch, H., Epple, M., Friess, W., Gelinsky, M., Greil, P., Hempel, U., Scharnweber, D., and Schulte, K. Functionally graded materials for biomedical applications. Mater Sci Eng A A362, 40, 2003.
3. Cuppone, M., Seedhom, B.B., Berry, E., and Ostell, A.E. The longitudinal young's modulus of cortical bone in the midshaft of human femur and its correlation with CT scanning data. Calcif Tissue Int 74, 302, 2004.
4. Bretcanu, O., Samaille, C., and Boccaccini, A.R. Simple methods to fabricate Bioglass-derived glass-ceramic scaffolds exhibiting porosity gradient. J Mater Sci 43, 4127, 2008.
5. Catledge, S.A., Clem, W.C., Shrikishen, N., Chowdhury, S., Stanishevsky, A.V., Koopman, M., and Vohra, Y.K. An electrospun triphasic nanofibrous scaffold for bone tissue engineering. Biomed Mater 2, 142, 2007.
6. Thomas, V., Zhang, X., Catledge, S.A., and Vohra, Y.K. Functionally graded electrospun scaffolds with tunable mechanical properties for vascular tissue regeneration, Biomed Mater 2, 224, 2007.
7. Salmoria, G.V., Ahrens, C.H., Klauss, P., Paggi, R.A., Oliveira, R.G., and Logo, A. Rapid manufacturing of polyethylene parts with controlled pore size gradients using selective sintering. Mat Res 10, 211, 2007.
8. Schaefer, D., Martin, I., Shastri, P., Padera, R.F., Langer, R., Freed, L.E., and Vunjak-Novakovic, G. In vitro generation of osteochondral composites. Biomaterials 21, 2599, 2000.
9. Gao, J., Dennis, J.E., Solchaga, L.A., Awadallah, A.S., Goldberg, V.M., and Caplan, A.I. Tissue engineered fabrication of an osteochondral composite graft using rat bone marrow-derived mesenchymal stem cells. Tissue Eng 7, 363, 2001.
10. Liao, S., Wang, W., Uo, M., Ohkawa, S., Akasaka, T., Tamura, K., Cui, F., and Watari, F. A three-layered nano-carbonated hydroxyapatite/collagen/PLGA composite membrane for guided tissue regeneration. Biomaterials 26, 7564, 2005.
11. Ghosh, S., Viana, J.C., Reis, R.L., and Mano, J.F. Bi-layered constructs based on poly(L-lactic acid) and starch for tissue engineering of osteochondral defects. Mater Sci Eng C 28, 80, 2008.

12. Sudarmadji, N., Tan, J.Y., Leong, K.F., Chua, C.K., and Loh, Y.T. Investigation of the mechanical properties and porosity relationships in selective laser-sintered polyhedral for functionally graded scaffolds. Acta Biomater 7, 530, 2011.
13. Tadmor Z. and C. Gogos, "Principles of polymer processing", Wiley, 2nd edition (2011).
14. Erisken, C., Kalyon, D.M., and Wang, H. A hybrid twin screw extrusion/electrospinning method to process nanoparticle-incorporated electrospun nanofibers. Nanotechnology 18, 1, 2008.
15. Erisken, C., Kalyon, D.M., and Wang, H. Functionally graded electrospun polycaprolactone and β-tricalcium phosphate nanocomposites for tissue engineering applications. Biomaterials 29, 4065, 2008.
16. Erisken, C., Kalyon, D.M., Wang, H., Ornek-Ballanco, C., and Xu, J. Osteochondral tissue formation through adipose-derived stromal cell differentiation on biomimetic polycaprolactone nanofibrous scaffolds with graded insulin and beta-glycerophosphate concentrations. Tissue Eng Part A 17, 1239, 2011.
17. Erisken, C., D. Kalyon and H. Wang, "Viscoelastic and biomechanical properties of tissue engineered constructs for osteochondral tissue regeneration", J. Biomechanical Engineering, 132, 091013-9 (2010).
18. Ozkan, S., D. Kalyon and X. Yu, "Multifunctional protein-encapsulated polycaprolactone scaffolds", NY Academy of Sciences, Chemical Engineering Approaches to Challenges in Energy and Biomedicine Meeting, March 30, 2012, New York, NY.
19. Ozkan, S., Kalyon, D.M., and Yu, X. Functionally graded β-TCP/PCL nanocomposite scaffolds for bone tissue engineering: In vitro evaluation with human fetal osteoblast cells. J Biomed Mater Res A 92, 1007, 2010.
20. S. Ozkan, D. M. Kalyon, X. Yu, C. McKelvey and M. Lowinger, "Multifunctional protein-encapsulated polycaprolactone scaffolds: Fabrication and in vitro assessment for tissue engineering", Biomaterials, 30, 4336-4347 (2009).
21. Ergun, A., Yu, X., Valdevit, A., Ritter, A., and Kalyon, D.M. In vitro analysis and mechanical properties of twin screw extruded single-layered and coextruded multilayered poly(caprolactone) scaffolds seeded with human fetal osteoblasts for bone tissue engineering. J Biomed Mater Res Part A 99A, 354 (2011).
22. Ergun A., D. Kalyon, A. Valdevit, A. Ritter , X. Yu, "Twin screw extrusion and co-extrusion based fabrication of uniform and multilayered scaffolds for bone tissue engineering", Orthopaedic Research Society, Transactions Vol. 36, 0285 (2011).
23. Ergun A., D. Kalyon, A. Valdevit and A. Ritter, "Compressive fatigue behavior of osteoblast seeded tissue constructs of poly(caprolactone) multilayered scaffolds for bone graft substitute applications", Orthopaedic Research Society, Transactions Vol. 36, 1850 (2011).
24. Ergun, A., Chung, R., Ward, D., Valdevit, A., Ritter, A. and Kalyon, D. M. Unitary bioresorbable cage/core bone graft substitutes for spinal arthrodesis coextruded from polycaprolactone biocomposites. Annals of Biomedical Engineering, 40, 1073, (2012).
25. Ergun, A., X. Yu, A. Valdevit, A. Ritter and D. M. Kalyon, "Radially and axially-graded multi-zonal scaffolds targeting critical-sized bone defects from polycaprolactone/hydroxyapatite/tricalcium phosphate", Tissue Engineering A (2012) doi:10.1089/ten.tea.2011.0625.
26. D. Gotsis, A. Ji and D. M. Kalyon, "3-D Analysis of the Flow in Co-rotating Twin Screw Extruders," Society of Plastics Engineers, ANTEC Technical Papers, 36, 139-142(1990).

27. Lawal A. and D. M. Kalyon, "Non-isothermal Model of Single Screw Extrusion of Generalized Newtonian Fluids," Numerical Heat Transfer, 26 (1), 103-121 (1994).
28. Lawal, A. and Kalyon D. M., "Non-Isothermal Extrusion Flow of Viscoplastic Fluids with Wall Slip", International Journal of Heat and Mass Transfer, 40 (1997) 3883-3897.
29. Lawal, A. and Kalyon D. M., "Viscous Heating in Nonisothermal Die Flows of Viscoplastic Fluids with Wall Slip," Chem. Eng. Sci., 52 (1997) 8, 1323-1337.
30. Kalyon D. M., A. Lawal, R. Yazici, P. Yaras and S. Railkar, "Mathematical Modeling and Experimental Studies of Twin Screw Extrusion of Filled Polymers", Polym. Eng. Sci., 39, 6, 1139-1151 (1999).
31. Lawal, A. and D. M. Kalyon, "Analysis of Nonisothermal Screw Extrusion Processing of Viscoplastic Fluids with Significant Backflow", Chem. Eng. Sci., 54, 999-1013 (1999).
32. Malik M. and Kalyon, D.M., "Three-dimensional Finite Element Simulation of Processing of Generalized Newtonian Fluids in Counter-rotating and Tangential Twin Screw Extruder and Die Combination", Int. Polym. Processing, 20 (2005) 398- 409.
33. Kalyon D. M., Apparent slip and viscoplasticity of concentrated suspensions, J. Rheol. 49 (2005) 621-640.
34. Kalyon D. M. and Malik M., "An integrated approach for numerical analysis of coupled flow and heat transfer in co-rotating twin screw extruders", International Polymer Processing, 22 (2007) 293-302.
35. Tang H. and Kalyon D. M., "Unsteady circular tube flow of compressible polymeric liquids subject to pressure-dependent wall slip", J. Rheology, 52 (2008) 507-526.
36. Tang H. S. and Kalyon D. M., "Time-dependent tube flow of compressible suspensions subject to pressure dependent wall slip: Ramifications on development of flow instabilities", Journal of Rheology 52 (2008) 5, 1069-1090.
37. Kalyon D.M. and Malik M., "Axial laminar flow of viscoplastic fluids in a concentric annulus subject to wall slip", Rheologica Acta (2012) 51:805–820.
38. Yilmazer U. and Kalyon D. M., Slip effects in capillary and parallel disk torsional flows of highly filled suspensions, J. Rheol. 33 (1989) 1197-1212.
39. Kalyon D. M. and Gokturk H. S., "Adjustable Gap Rheometer," U.S. Patent and Trademark Office # 5,277,058, Jan. 11, 1994.
40. Aral B. and Kalyon D. M., Effects of temperature and surface roughness on time-dependent development of wall slip in steady torsional flow of concentrated suspensions, J. Rheol. 38 (1994) 957-972.
41. Lawal, A., Kalyon D. M., Squeeze flow of viscoplastic fluids subject to wall slip, Polymer Eng. Sci. 38 (1998) 1793-1804.
42. Tang H. and Kalyon D. M., "Estimation of the Parameters of Herschel-Bulkley Fluid under Wall Slip Using a Combination of Capillary and Squeeze Flow Viscometers", Rheologica Acta, 43 (2004) 1 80-88.
43. Kalyon D. M. and Tang H., "Inverse problem solution of squeeze flow for parameters of generalized Newtonian fluid and wall slip", Journal of Non-Newtonian Fluid Mechanics, 143 (2007) 133-140.

(249c) Investigating Angiogenesis in Living Vascular Networks in Vitro

John P. Morgan, Chemical and Biomolecular Engineering, Cornell University, Ithaca, NY

Investigating Angiogenesis in Living Vascular Networks *In Vitro*

Microvascular networks support metabolic activity and define the microenvironmental conditions within tissues during healthy and pathological processes, and provide the foundation for embryonic development. Following its *de novo* synthesis, the microvasculature develops and remodels via the process of angiogenesis, the formation of new capillaries from preexisting blood vessels by the sprouting or splitting of preexisting vessels into a hierarchically branching vascular network. In this presentation, we describe how we are studying angiogenesis by engineering living vascular networks *in vitro* in a microfluidic culture, using lithographic techniques to form endothelialized microvessels within a native collagen matrix. The role of physical and chemical forces in angiogenesis is investigated experimentally and modeled mathematically.

Angiogenesis is a tightly controlled sequence of events involving endothelial cell sprouting, migration and proliferation; degradation of tissue; new capillary vessel formation and loop formation (anastomosis). However, none of these events occur until the onset of flow, with sufficient viscosity and hence shear stress, through the network. We hypothesize that mechanical stresses provide the key triggering event to initiate biochemical signaling pathways that guide subsequent growth. Further, these mechanically induced changes evolve both spatially and temporally in response to complimentary and competing effects of chemotaxis and perfusion-related hemodynamic forces. Post-development, angiogenesis also occurs during wound healing, menstruation, in response to hypoxia (low oxygen levels), and in pathologies such as cancer. Elucidation of the basic mechanism of angiogenesis and related triggering stimuli could, for example, lead to the clarification of mechanobiological signaling pathways, aid in the refinement of anti-angiogenic cancer therapies and form the basis of strategies to engineer vascularized tissues for regenerative medicine.

We will present our work toward the formation of functional microvascular structures *in vitro*. We will describe our strategies for the formation of endothelialized vessels within cell-remodelable matrices and our experimental methods for defining physiologically relevant perfusion stresses and biochemical gradients during extended cultures. We will then turn to a discussion of our observations of the response of these vessels, as evaluated via live microscopy and at immunohistochemically stained end points. We will focus on characterization of morphology, and long-term stability of endothelia, as a function of individual physical forces (e.g., shear stress and wall tension) and different flow regimes (e.g., pulsatile, laminar, and oscillatory). We will further discuss the behavior of co-cultures with perivascular cells in the bulk of the matrix. We will conclude with a discussion of how our observations relate to observations in *in vivo* studies and how they inform mathematical models of angiogenic maturation.

USE OF LATE EMBRYOGENESIS PROTEINS TO ENGINEER DESICCATION TOLERANCE IN MAMMALIAN CELLS

Nilay Chakraborty[1,2], Shumin Li[3*], Apurva Borcar[3], Steven C. Hand[3], Mehmet Toner[2]*

[1]University of Michigan-Dearborn, Dearborn, MI, USA
[2]Harvard Medical School and Massachusetts General Hospital, Boston, MA, USA
[3]Louisiana State University, Baton Rouge, LA, USA

Introduction

The mechanisms by which different anhydrobiotic organisms protect their cellular structure and function during desiccation can be an important strategy for cell stabilization to support cell-based therapies. While it is known that sugars such as trehalose can help in stabilizing biological structures during desiccation (1, 2), recent studies underscore the importance of various protective proteins during drying that include Late Embryogenesis Abundant (LEA) proteins (3, 4). Late embryogenesis abundant (LEA) proteins are associated with desiccation tolerance in several anhydrobiotic organisms including insects, tardigrades and rotifers (3, 4).

The objective of this study was to use the biomimetic strategy of using LEA proteins to engineer desiccation tolerance in mammalian cells for lyopreservation. We have identified and sequenced two LEA genes (*Afrlea2* and *Afrlea3m*) in the embryo of desiccation tolerant brine shrimps (*Artemia franciscana*). AfrLEA3m was shown to be localized in mitochondria while AfrLEA2 is cytoplasmic (REF). In this study, human hepatocellular carcinoma (HepG2) cells with AfrLEA2 and AfrLEA3m was used to evaluate the possibility of improved survivorship during drying. The technique of spin-drying was used to rapidly and uniformly desiccates cells to a moisture content of 0.2 gH2O/gdw at 22^{0}C. We report dramatic improvement of desiccation tolerance of HepG2 cells using AfrLEA3m proteins. Desiccated cells showed significantly higher survival after drying and rehydration.

Materials and Methods

Human hepatocellular carcinoma (HepG2) cells were stably transfected to express AfrLEA2 and AfrLEA3m using a tetracycline-inducible expression system. Trehalose was loaded into these cells using a TRET1 trehalose transporter isolated from anhydrobiotic larvae of Polypedilum vanderplanki (REF kikawada). The expression vector pcDNA 6.2/V5-DEST (Gateway Vector, Invitrogen) was used for constitutive expression of the TRET1 trehalose transporter (6). Transfection of HepG2 Tet-On cell lines (with or without the capacity for AfrLEA2/AfrLEA3m expression) was accomplished with Geneporter transfection reagent (Genlantis, San Diego, CA, USA). Stable cell lines were selected using blasticidine.

Monolayers of HepG2 cells expressing various combinations of AfrLEA2, AfrLEA3m and TRET1 were allowed to attach to 22 mm collagen coated round glass coverslips (BD BioCoat, San Jose, CA) by incubating 18 h in cell culture medium. For all cells expressing TRET1, the medium also contained 50 mM trehalose. To prevent cells from attaching to the edge of the coverslips, 20 mm diameter Press-to-Seal™ silicone isolators (Invitrogen, Carlsbad, CA) were used. Immediately before spin-drying, the cell culture media was removed completely using a Pasteur pipette and replaced with the spin-drying solution (1.8 M trehalose, 10 mM KCl, 5 mM glucose, 20 mM HEPES, and 120 mM choline chloride at pH 7.4). This trehalose-containing solution was essential for successful cellular drying outcomes regardless of the intracellular variables tested to avoid extensive cellular injury due to ionic imbalance. Spin-drying was performed using a commercially available spin-coater (Brewer Science Inc. St. Louis, MO). The basic spin-drying process is described in the Figure 1. Cells attached to glass coverslips were spin dried at 1000 rpm for 60 sec in a trehalose buffer at room temperature. Following spin-drying, the samples were immediately rehydrated and the membrane integrity of the cells following desiccation was assessed using a Syto-13/Ethidium bromide fluorescence assay. Desiccated cells were cultured under normal tissue culture conditions following rehydration for 7 days to evaluate long-term growth characteristics.

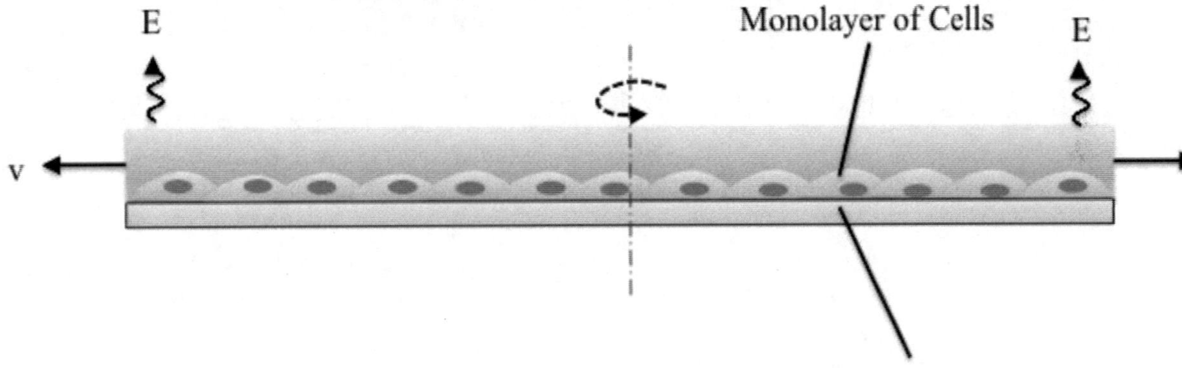

Figure 1: Schematic explaining the technique of spin-drying of cells. The arrows indicate different mechanism of moisture removal during spin-drying. At the beginning of the spin-drying process the excess liquid is removed from the sample by radial flow (v). After the excess liquid is removed by radial flow, the evaporation (E) into an environment continuously purged with dry nitrogen gas causes further drying of the samples.

Results and Discussion

Control HepG2 cells (without LEA protein or intracellular trehalose) exhibited 0 % membrane integrity (n = 9). HepG2-TRET1 cells preloaded with trehalose showed 44.5 ± 22.2 % integrity (mean ± SD, n = 3); HepG2-AfrLEA2 cells without trehalose, 57.2 ± 13.0 % (n = 9) (Figure 2); HepG2-AfrLEA2 cells preloaded with trehalose, 98.3 ± 2.2 % (n = 9); HepG2-AfrLEA3 cells without trehalose, 93.6 ± 4.6 % (n = 9); and HepG2-AfrLEA3 preloaded with trehalose, 97.7 ± 3.8 % (n = 9). Growth studies across subsequent days after rehydration revealed higher proliferation for cells containing LEA protein than those without. Between AfrLEA2 (cytoplasmic) and AfrLEA3m (mitochondrial) proteins, AfrLEA3m was able to provide better protection against desiccation stresses as evidenced by superior proliferation rates. Growth studies also demonstrate that the LEA proteins act synergistically with trehalose in providing protection against desiccation stresses and adding intracellular trehalose along with LEA proteins enhances the long-term growth of cells, as measured across 7 days post-rehydration, to proliferation rates that approach those of control cells never dried.

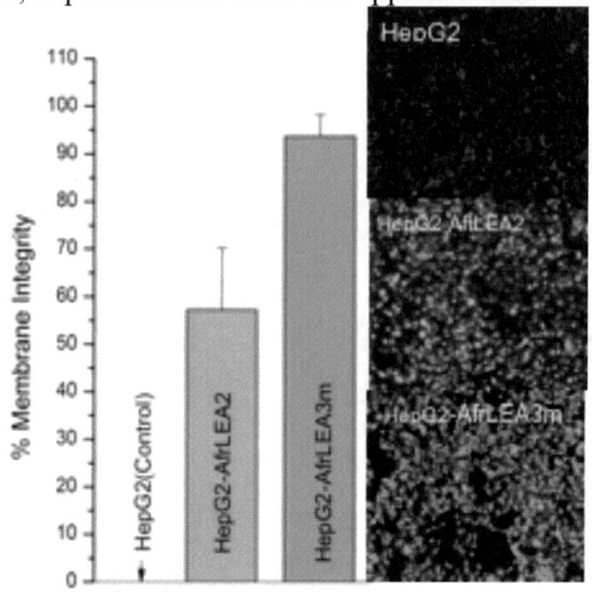

Figure 2. Membrane integrity of the HepG2 cells expressing Late Embryogenesis Proteins. Fluorescent micrographs of the rehydrated HepG2 cells indicate the protective of the LEA proteins.

Conclusion

This study outlines a biomimetic approach of using LEA proteins to dramatically enhance desiccation tolerance in mammalian cells. LEA proteins offer an excellent opportunity for engineering cellular biostability in the dried state.

References

1. Crowe JH., Crowe LM., Wolkers WF., Oliver AE., Ma X, Auh J-H, Tang M, Zhu S, Norris J, and Tablin F. (2005), "Stabilization of dry mammalian cells: Lessons from nature," *Integr Comp Biol,* 45(5):810-820.
2. Xie G, Timasheff SN (1997), "The thermodynamic mechanism of protein stabilization by trehalose," *Biophys Chem* 64(1-3):25-43.
3. Tunnacliffe A and Wise MJ (2007), "The continuing conundrum of the LEA proteins," *Naturwissenschaften* 94(10):791-812.
4. Hand SC, Menze MA, Toner M, Boswell L and Moore D (2011), "LEA proteins during water stress: Not just for plants anymore," *Annu Rev Physiol* 73(1):115-134.
5. T. Kikawada, A. Saito, Y. Kanamori, Y. Nakahara, K. Iwata, D. Tanaka, M. Watanabe, T. Okuda (2007), "Trehalose transporter 1, a facilitated and high-capacity trehalose transporter, allows exogenous trehalose uptake into cells," *Proc. Natl. Acad. Sci. USA* 104:11585–11590.
6. Chakraborty N**,** Menze M, Hand SC and Toner M (2012) Trehalose Transporter from African Chironomid Larvae Improves Desiccation Tolerance of Chinese Hamster Ovary Cells. *Cryobiology* 64:91-96.

(249f) Hemoglobin Regulates the Migration of Glioma Cells Along Poly(ε-caprolactone)-Aligned Nanofibers

Alexander Roth[1], Ruipeng Xue[2], Tyler Nelson[3], Jed Johnson[4], Jacob Elmer[1], Joseph Huntley[1], John J. Lannutti[5], Mariano S. Viapiano[6] and Andre Palmer[1], (1)Department of Chemical and Biomolecular Engineering, The Ohio State University, Columbus, OH, (2)Materials Science and Engineering, The Ohio State University, Columbus, OH, (3)Department of Biomedical Engineering, The Ohio State University, Columbus , OH, (4)Material Science and Engineering, Ohio State University, Columbus, OH, (5)Department of Material Science and Engineering, The Ohio State University, Columbus, OH, (6)Neurological Surgery, The Ohio State University, Columbus, OH

This work seeks to integrate two disparate technologies and examine their combined effect on U251 glioma cell migration. Aligned fibers have been shown to facilitate cell migration in the direction of fiber alignment while oxygen-carrying solutions improve the metabolism of cells in hypoxic culture. Therefore in this work, U251 aggregate migration on poly(ε-caprolactone) (PCL) aligned fibers was studied in cell culture media supplemented with the O_2 storage and transport protein hemoglobin (Hb) obtained from bovine, earthworm and human sources at concentrations ranging from 0-5 g/L within a cell culture incubator and oxygen tensions ranging from 1-19% O_2. In one set of experiments conducted on tissue culture polystyrene – representing a surface conducive to random cell movement – cell viability and the total number of live cells were compared for individually cultured U251 cells under different culture conditions. To determine the effect of Hb concentration and incubator O_2 tension on Hb oxidation, Hb auto-oxidation rate constants were measured 24 hours after Hb incubation at 37°C. Individual cell migration was also quantified using a wound healing assay up to 24 hours after cell seeding. In another set of experiments, U251 cell aggregates were developed and aggregate dispersion/cell migration quantified for cells cultured on PCL-aligned fibers up to 72 hours after aggregate seeding. The results of this work show that the presence of bovine or earthworm Hb improved individual cell viability at 1% O_2 while human Hb adversely affected cell viability at increasing Hb concentrations and decreasing oxygen levels. It was also observed that Hb auto-oxidation rate constants increased with decreasing O_2 levels, increasing Hb concentrations, and was greatest for bovine Hb and the least for earthworm Hb under identical culture conditions. The presence of cells decreased the rate of Hb auto-oxidation. The wound healing assay showed no differences in individual cell migration under most conditions, except for an increase in cell migration at 0.5 g/L bovine Hb at 5% O_2, and a reduction in cell migration at 1% O_2 in the presence of 5 g/L earthworm Hb. The control data also suggests that decreasing the O_2 tension in the incubator from 5% O_2 to 1% O_2 decreased aggregate dispersion on the PCL aligned fibers. However, the addition of bovine Hb at 5% O_2 significantly improved aggregate dispersion. At 19% O_2 Hb did not impact aggregate dispersion. Also at 1% O_2, aggregate dispersion appeared to increase in the presence of earthworm Hb, but only at the latter time points. Taken together, these results show that Hb-based O_2 carriers can be utilized to improve O_2 availability and the migration of glioma spheroids *in vitro*.

(262a) Protein Behavior At Pharmaceutically-Relevant Microparticle Interfaces

Theodore W. Randolph, Department of Chemical and Biological Engineering, Center for Pharmaceutical Biotechnology, University of Colorado, Boulder, CO

During shipping, storage and delivery to patients, therapeutic proteins are exposed to a number of interfaces, to which they readily adsorb. These interfaces include fluid-fluid interfaces such as that presented by microdroplets of emulsified silicone oil used to lubricate prefilled syringes, as well a as solid-liquid interfaces such as those presented by contaminants such as glass microparticles that slough off vial walls and steel microparticles shed by high-speed filling pumps. Adsorption to these interfaces may result in conformational changes in the adsorbed proteins, and may also lead to protein aggregation and the formation of large particles that render a product unacceptable for use in humans. In addition, adsorption to a microparticle may substantially increase that a therapeutic protein will provoke an undesirable immune response. The adsorption behavior of several therapeutic proteins on a number of these interfaces will be described, along with measurements of tertiary structural changes in the proteins that can be detected by front-face fluorescence quenching measurements. Finally, the results of tests in murine models to determine the effect of such particles on immune response are described.

(262b) Single Molecule Protein Dynamics and Aggregation At the Oil-Water Interface

Robert Walder and Daniel K. Schwartz, Chemical and Biological Engineering, University of Colorado at Boulder, Boulder, CO

The dynamic behavior of proteins at liquid-liquid interfaces is of fundamental interest to many applications, including complex fluids formulation, biosensing, and pharmaceutical processing and delivery. However, the methods for understanding proteins at fluid interfaces have been limited to a small number of traditional methods (such as interfacial rheology), because of the experimental complexities associated with investigating fluid interfaces. We have recently developed methods based on single molecule total internal reflectance fluorescence (smTIRF) microscopy to investigate the dynamics and aggregation of proteins at the silicone oil-water interface. Using high throughput single molecule tracking, we track each individual protein from adsorption to the interface, through diffusive interfacial transport and finally till desorption back into the bulk water phase. We have discovered a rich diversity of dynamic phenomena, including reversible adsorption, diffusion and desorption correlated with aggregate state, and multiple modes of diffusion. We have also measured the nonlinear kinetics associated with interfacial protein layer formation and modeled this process using a kinetic population balance model. Importantly, this model includes desorption since we explicit observe it in our single molecule tracking experiments. In this talk, I will show how using this single molecule approach has given us these new insights, many of which could not be observed using traditional ensemble averaged methods.

(262c) Does Poly(ethylene glycol) Conjugation Protect Proteins From Conformational Change During Emulsion-Based Microsphere Encapsulation?

Adam L. Canady, Chemical Engineering, Carnegie Mellon University, Pittsburgh, PA, Todd M. Przybycien, Department of Chemical Engineering and Department of Biomedical Engineering, Carnegie Mellon University, Pittsburgh, PA and Robert D. Tilton, Chemical Engineering and Biomedical Engineering, Carnegie Mellon University, Pittsburgh, PA

Due to their high activity and specificity, proteins are prime candidates for use as therapeutic drugs. Since proteins cannot be administered orally, alternative delivery strategies are required. The typical route is via injection, which leads to fluctuations in the concentration of drug in serum and often is associated with poor patient compliance. To improve efficacy in chronic therapies, protein drugs would benefit from sustained release delivery. One potential candidate would be a biodegradable poly(lactide-co-glycolide) (PLG) depot which often takes the form of subcutaneously injected microspheres. Despite its benefits, this delivery method presents its own challenges, namely that proteins tend to adsorb to the interfaces generated during the formation of and release from microspheres. Adsorption can be irreversible and result in protein denaturation, aggregation and, ultimately, bioactivity loss. The covalent attachment of poly(ethylene glycol) (PEG) chains to the protein ("PEGylation") has been shown to mitigate adsorption effects in addition to imbuing conjugates with several other useful benefits. PEGylated proteins tend to better preserve bioactivity during sustained release and to experience more complete release from PLG microspheres. Prior research has shown that PEGylation significantly modulates protein adsorption isotherms, adsorption reversibility and surface-induced aggregation at the solid PLG/aqueous interface, which is relevant to the release from PLG depots. The current work focuses on adsorption at the oil/water interface, which is relevant to possible protein damage during the emulsion-based microsphere formation process. Understanding how PEGylation affects protein adsorption to the oil/water interface formed during microsphere preparation will facilitate improved formulations for PLG depot-based protein therapeutics. To this end, two model proteins, ribonuclease A (RNase A) and apo-α-lactalbumin (ALA), were grafted with 20kDa PEG chains. RNase A and ALA differ substantially in their thermodynamic stability and hydrophobicity. Circular dichroism and fluorescence spectroscopy were used to elucidate how PEGylation alters the conformational effects of protein exposure to the oil/water interface during emulsification. RNase A suffered little conformational alteration during oil/water interface exposure under emulsification conditions, while ALA experienced significant changes in tertiary structure. PEGylation proved to have no protective effect against conformational change at the oil/water interface for either protein. Differences in emulsification-induced conformational change are being interpreted with the aid of ellipsometry measurements of the extent of adsorption and interfacial tension measurements for the oil/water interface, for both PEGylated and unmodified proteins. The tentative conclusion is that the benefit of protein PEGylation in terms of enhanced bioactivity preservation and increased extent of release from PLG microspheres arises more from its effects on adsorption to the solid PLG/aqueous interface than to protection during microsphere encapsulation.

(262d) Structure and Stability of Pol(L-lysine)/Hyaluronan Thin Films As Nanoreservoirs for the Bone Morphogenetic Protein-2

Flora Gilde, Ofelia Maniti, Raphael Guillot, **Jorge Almodovar** and Catherine Picart, Grenoble Institute of Technology, Grenoble, France

Polyelectrolyte multilayer films (PEM) made of poly(L-lysine) and hyaluronan (PLL/HA) have been recently investigated for their ability to trap the osteoinductive recombinant human bone morphogenetic protein-2 (rhBMP-2) (Crouzier et al., 2009). However, several tests have to be performed prior to their possible use in clinics as bioactive coatings on orthopedic or dental implants. First, all medical devices should sustain a sterilization process before *in vivo* implantation and they should also sustain storage for at least several months. In view of future applications of the PEM films, we first aimed here to investigate the stability, shelf life-time and sterilization of (PLL/HA) film. Secondly we aimed to investigate the secondary structure of rhBMP-2 trapped hydrated or dry (PLL/HA) films, as compared to the protein in solution at different pHs. To this end, we used Fourier Transform Infrared Spectroscopy (FTIR) to investigate the structure of (PLL/HA) films in dry state, their stability over time, the effects of sterilization by gamma irradiation and the structure of rhBMP-2 trapped in the films. We found that the (PLL/HA) films were very stable over the 2-month period, with less than 2% variation in the absorbance intensity of the amide I band. We also proved that the film can withstand the sterilization process without any major change in their structure. We quantified the secondary structure of rhBMP-2 trapped in a hydrated film and observed that it exhibited a similar structure to that of the protein in solution at pH 3 (representing the optimal conditions for BMP-2 loading the film). The preservation of the secondary structure of the film when trapped and confined in the PEM film may explain why rhBMP-2 remained bioactive. Interestingly, when the protein was trapped in dry films, a significant change in the protein structure was visible with the appearance of intermolecular b-sheets. We attributed these changes to protein/protein interactions or to protein/PLL interactions.

Crouzier, T., K. Ren, C. Nicolas, C. Roy, and C. Picart. 2009. Layer-by-Layer films as a biomimetic reservoir for rhBMP-2 delivery: controlled differentiation of myoblasts to osteoblasts. *Small*. 5:598-608.

(262e) Formulation of Bovine Serum Albumin Encapsulated Cal-Alginate Microspheres by Electrodispersion for a Simulated Release in a Gastric Fluid

Yinyan Zhao, Bioanalytical Systems, Inc, West Lafayette, IN, Fuyue Li, Chemical Engineering, Georgia Institute of Technology, Atlanta, GA and **Michael T. Harris**, School of Chemical Engineering, Purdue University, West Lafayette, IN

Calcium alginate (Ca-alginate) gels have been widely used for encapsulation of variety of biological active agents, including proteins, peptides and cells due to the relatively mild cross-linking conditions of preparation. However, most research studies report using the common technique, the droplet dripping method without size control (>500μm). The purpose of this study is to evaluate the new application of the electrodispersion technique for the formulation of Ca-alginate microcapsules suitable for biomedical applications, and assess the effect of the production methods on the bioactivity of protein to ensure that the encapsulated protein is able to preserve its stability during the particle processing and recovery. The model protein, bovine serum albumin (BSA) was used for this study. Furthermore, the release profile of BSA in the stimulated gastrointestinal conditions was investigated. The mean diameter of BSA loaded gel microspheres was measured to be less than 20μm and an encapsulation efficiency of nearly 65% was achieved. BSA from Ca-alginate particulate system demonstrates a biphasic release profile in a simulated GI tract. The secondary structure of BSA was monitored with and without alginate matrix. A structural integrity of BSA was observed before and after the process. The study shows that the electrodispersion technique is feasible to produce BSA encapsulated Ca-alginate gel microspheres intended for oral administration.

Abstract
Design of a Vessel to Facilitate Rapid Thawing of a Labile Protein

Bob Large and Ajoy Velayudhan
Process Development and Technology
Grifols, Inc.
Clayton, North Carolina

Human blood plasma contains several proteins of therapeutic interest. From the time of acquisition from the donor until separation and purification at the manufacturing site, the plasma is maintained frozen to preserve protein functionality. The first step of the separation process requires the gentle thawing of a large pool of frozen plasma. Previous studies have shown that minimizing the thaw time increases the recovery and potency of key proteins. Effective mixing is essential to reduce thaw times; however, mixing must be achieved with low local shear rates, since key proteins are known to be sensitive to shear.

Effective mixing of the frozen plasma under conditions of low shear was demonstrated in a horizontal thawing vessel, thereby facilitating rapid thawing with minimal loss of protein potency. Operation at the bench and pilot scales have shown the horizontal configuration capable of an increase in throughput of 20% and select protein recovery increase of at least 35%.

Human blood plasma contains multiple proteins of therapeutic interest. Purification processes have been developed to separate out individual key proteins. These individual proteins are then used to treat different diseases in humans. Two primary reasons exist for individuals to need these therapies. In some cases the proteins provide replacement for individuals that have experienced trauma, e.g., albumin, or are undergoing chemotherapy, e.g., immune globulins. In other instances they are delivered to those with a deficiency in the body's ability to produce the protein, e.g., clotting factors for hemophilia or immune globulins for immunodeficiency. Plasma is collected from healthy individuals at donation centers and immediately frozen to maintain protein functionality while being transferred to the manufacturing facility.

The first step of the separation process requires the gentle thawing of several thousand combined plasma units.

Proteins are very large and complex molecules that need to be handled carefully. Protein functionality may be lost by excessive levels of factors such as temperature or shear stress. These factors require the thawing process to be quick, gentle and at the lowest temperature that is feasible. Past studies have shown that minimizing the thaw time increases the recovery and potency of key proteins. Mixing during the thawing process is necessary to support heat transfer; however, the mixing must be designed to reduce mechanical stress on labile proteins.

Minimizing time, system temperature and mechanical stress creates a need to balance opposing forces. Minimizing time while keeping the solution near freezing can be done by maximizing the jacket temperature. However, maximizing the jacket temperature with slow mixing to minimize stress creates the opportunity for hot spots in the solution. One current approach is a vertical vessel designed with a slow moving agitator. The price to be paid in this configuration is that the less dense still frozen material floats to the surface. Plasma-ice floating on the surface, partially exposed to the air and liquid, has reduced heat transfer from the already thawed liquid in contact with the heated jacket. As batch size increases another issue with the vertical design becomes apparent. To maximize surface area to volume ratio requires minimizing the vessel diameter. The requirement then leads to the need to increase the height of the vessel to maintain operating volume. The result is an increase in the thawed volume that accumulates below the floating ice layer. The separate ice/liquid layers leads to temperature control challenges. This is an effective approach; however, another approach has proven to decrease thaw time while keeping the solution temperature near freezing resulting in increased protein recovery.

To balance the opposing forces, a vessel of horizontal configuration was investigated. Typically this configuration is utilized to facilitate thorough mixing of powders at high speeds or evaporation of volatile compounds from solids. These processes are dealing with a stable solid that can handle both high speed mixing and high temperatures. As a novel approach to protein thawing, a horizontal configuration was considered for gentle thawing of labile proteins.

The main advantage of the horizontal configuration is that a properly designed agitator will push the frozen material under the liquid level, even at low agitation rates. Thorough mixing of the frozen and thawed material in this manner will eliminate stratification in the vessel regardless of any increase in operating volume. The increased solid/liquid contact will enhance heat transfer, reduce thawing time, and reduce/minimize liquid hot spots. Further understanding was achieved by investigating the horizontal design at three scales; designated herein based on increasing size as Scale 1, Scale 2, and Scale 3. To compare the current vertical design to the horizontal approach several characteristics were monitored. Of those the most significant were determined to be thawing time and potency of a particularly labile protein which will be designated as Protein Y.

To determine feasibility a vessel at Scale 1, a bench top unit, was constructed of clear acrylic. The trajectories of the frozen blocks could be seen, and due to the size the unit could be operated vertically or horizontally. The ability to operate in both configurations allows for direct comparison of results. At Scale 1 data was gathered at combinations of two fill levels (40% and 100%) and two jacket temperatures (20ºC and 30ºC) for both vertical and horizontal placement. Trials were executed using water-ice initially to identify thaw time differences. Next, proportionally sized plasma blocks were used for comparison of protein recovery to the current manufacturing vertical vessel.

The two fill levels were chosen to compare the volume to surface area influence for the two configurations to thaw time. Below half full a horizontal vessel has a higher surface area to volume ratio than an equivalent volume in a similarly dimensioned vertical vessel, of course this is reversed above the 50% level. For ease of comparison the time was determined based on the solution reaching 2ºC. Using only water-ice and a jacket temperature of 20ºC, the horizontal position at the 40% fill level reduced the thaw time by 50%

and at a 100% fill the horizontal time was 20% less. Clearly the horizontal configuration is superior to the vertical position for efficiency of heat transfer, even at 100% where surface area:volume is equivalent for both configurations. These results highlight the substantial limitations of the vertical configuration.

Next the effect on Protein Y potency by the two configurations was investigated. Plasma is typically collected in 1 L bottles. To provide comparison to current manufacturing the bottles of plasma were cut into blocks to provide a feed of proportional size to manufacturing for use in the Scale 1 vessel. Filling the vessel with plasma blocks resulted in a liquid level of approximately 60% volume once thawed. Trials were conducted in both horizontal and vertical orientation with each position repeated at a jacket setting of 20°C and 30°C.

Similar reductions in thaw time were seen with plasma; horizontal operation was 45% faster than vertical for both jacket temperature settings. At 30°C the thaw time was an additional 35% faster in both orientations when compared to the jacket at 20°C. The reductions in thaw time translated directly to increases in Protein Y potency.

Protein Y potency data was captured for all combinations of horizontal and vertical orientation and jacket temperatures of both 20°C and 30°C. The lowest recovery was experienced in the vertical configuration with a jacket of 20°C. Taking this point to be represented by a potency of X, Table 1 provides a relative comparison of the results for all four configurations.

Table 1. Protein Y relative potency (Scale1)

	Vertical vessel	Horizontal vessel
Jacket 20°C	X	3X – 5X
Jacket 30°C	2X	3X – 7X

Considering the scale of the unit and the wide scatter in the potency results, the best conclusion is that the reduced thaw time for the horizontal configuration is beneficial. Past investigations of the impact from reducing the thaw time indicated a potential Protein Y potency gain of 20%. This evidence provided the justification to expand the trials to Scale 2.

The larger Scale 2 studies were conducted in a specially constructed horizontal vessel with strong consideration of the intended use. In this case, due to size and weight, the vessel could not be oriented in a vertical position. However, another study of the same protein using a similar sized vertical vessel was underway. These two vessels provided the comparator data for Scale 2.

Scale 1 showed a variation in efficiency based on operating level. Before beginning trials as Scale 2 an appropriate operating level was investigated. The horizontal vessel vendor indicated the optimal level is typically 55% to 60% for mixing operations. Tests were executed from 50% to 65% at Scale 2 with the horizontal vessel to understand operating level on heat transfer efficiency. A level of 60% was shown to provide the shortest cycle time without impacting other operating conditions (foaming, protein quality, product temperature, etc.).

The trials were conducted in both vessels at the 60% fill volume and a jacket temperature of 20°C. The horizontal vessel was able to thaw the plasma with a 65% reduction in time compared to the vertical unit. The higher relative efficiency when compared to Scale 1 may be attributable to the Scale 2 horizontal vessel being constructed by a vessel manufacturer knowledgeable in this configuration. Where the Scale 1 unit was constructed of materials found in the lab to determine design feasibility.

Both configurations experienced an increase in potency (Table 2), referenced to Scale 1 vertical vessel as defined above. The relative recovery difference from vertical to horizontal is slightly less at Scale 2. However, the recovery for the vertical vessel currently used in manufacturing (larger volume than Scale 3 horizontal vessel) is similar to the Scale 2 vertical vessel, no further benefit is experienced with increased scale for vertical vessels. To date horizontal vessel potency data is not available for greater than Scale 2. If as seen in the vertical configuration the horizontal recovery stays flat, from Scale 2 to Scale 3, there is still a significant advantage to the design.

Table 2. Protein Y relative potency (Scale1 and Scale 2 at 20°C)

	Vertical vessel	Horizontal vessel
Scale 1	X	3X – 5X
Scale 2	4X	10X

To understand the opportunity for generation of hot spots in the vessel a temperature mapping exercise was executed. A sensor that could survive the environment when ice was present could not be found; self contained units were damaged by the ice and thermocouples taped to the vessel wall were pulled free by the passing ice blocks. The vessel was filled to 60% with water only and four thermocouples were taped to the interior of the drum, Figure 1. Thermocouple T-2 was placed adjacent to the vessel's integral Resistance Temperature Device (RTD) (RTD-02 in Figure 2) for comparison of contents temperature to that indicated for the vessel. Placement of thermocouples T-1 and T-3 are at the bottom front and back corners of the vessel, while T-4 was placed beside the mixer shaft below the liquid surface. These areas were identified as the worst case positions for generation of mixing dead zones that could lead to hot spots. Figure 2 shows that changes in the jacket temperature are quickly seen by all the contents in the vessel, as indicated by the overlap of all thermocouple and RTD curves in the saw tooth pattern. More importantly, the temperature change throughout the vessel is homogeneous as a result of the mixing regime.

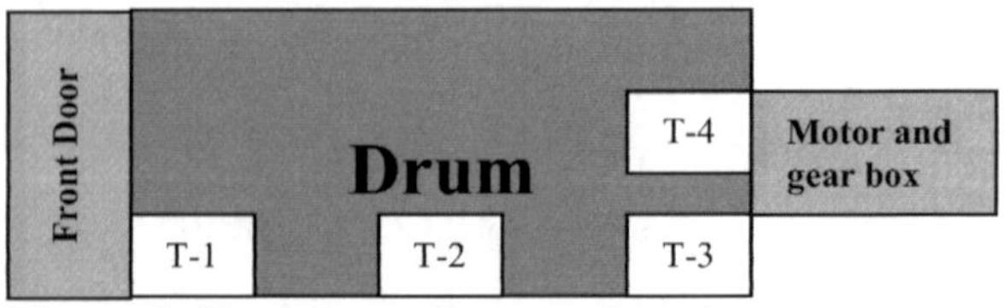

Figure 1: Temperature sensor locations

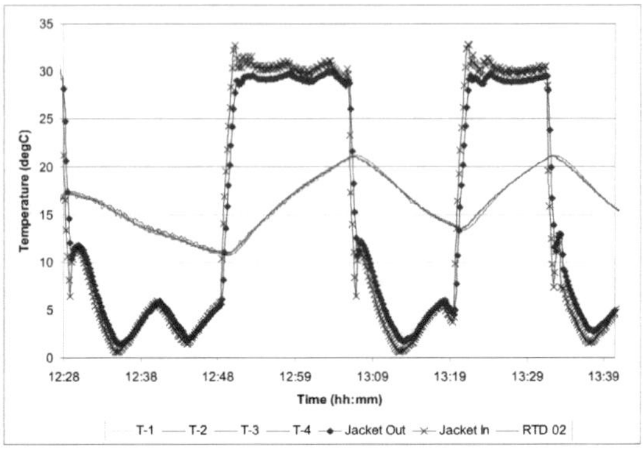

Figure 2: Temperature sensor results

Based on the efficiencies at Scales 1 and 2 a larger unit was constructed at Scale 3. One approach, common in bioreactor scale-up[1,2] is to evaluate the various relevant dimensionless parameters, and to characterize the dominant process at various scales. Knowing the standard sizes that are typically manufactured by the vendor a vessel of expected capacity was evaluated based on the given diameter. From that we can develop constants used to evaluate the scale-up parameters by holding certain criteria unchanged. In this case Energy Input divided by Volume (Po/V), Agitation Rate (N),

Table 3: Dimensionless scaling constants

Scale-up Criterion	Equation	Scale 2	Scale 3 Scale-up Estimates Po/V	N	N*D	Fr	Scale 3 Operation N*D	Scale 3 Test N*D
Energy Input/Volume	Po / V	**6.8**	**6.8**	33.6	3.1	10.2	2.5	1.9
Agitation Rate	N	**30**	18	**30**	14	20	14	10
Max. Impeller Speed	N*D	**15**	19.6	33.3	**15**	22.3	16.1	11.9
Froude (dimensionless)	N^2D/g	**45.9**	35.2	101.9	20.7	**45.9**	22.1	12.1

Maximum Impeller Speed (N*D) and the dimensionless Froude number. Others were determined to increase process understanding; however these four were focused on to determine appropriateness of size and operating conditions. The constant power input (Po/V) and constant shear at the impeller tip (N*D) were chosen to minimize protein loss to shear stress. The Froude number (Fr) is used in the scale-up of mixing processes in which the goal is to entrain a gas in a liquid. The aim here is to minimize or eliminate product loss from foaming (gas entrainment) which leads to the desire to minimize the Froude number. With all Scale 2 criteria being held at 1.0, Table 3 was developed to provide scaling constants to Scale 3; only computed results shown, scaling factors are not indicated.

Review of the Scale 3 Scale-up Estimates in Table 3 indicates the desired approach will be holding the tip speed (N*D) constant to minimize scale-up criteria. The delivered unit was operated at the estimated agitation rate (13 rpm) and required slightly more power than estimated. Because of a larger relative volume of the delivered unit the Po/V term was smaller; however, the tip speed and Froude number were larger.

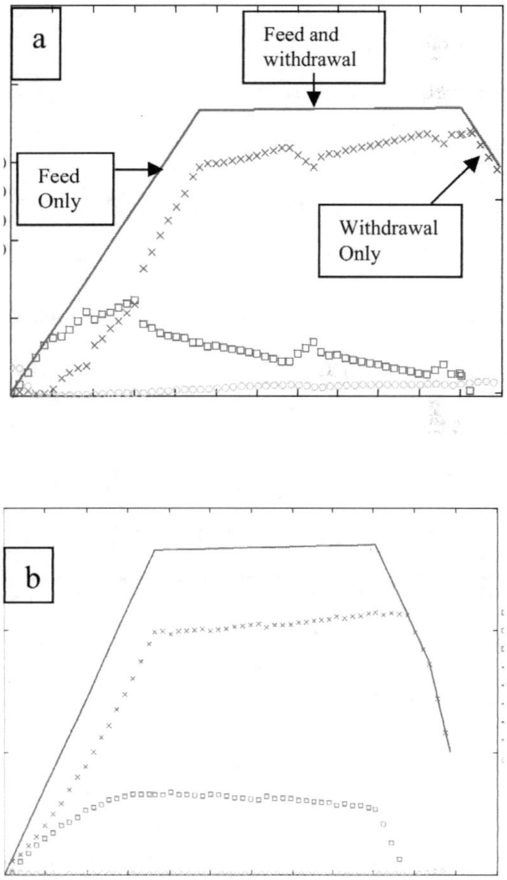

Figure 3: Operating level profile Scale 2 (frame a) and 3 (frame b); y-axis level, x-axis time, solid line liquid and ice total, x is liquid level, square is ice level, circle is contents temperature

A trial was run with a reduced agitation rate at 10 rpm, last column of Table 3. The three scaling factors each are below the estimates, thus reducing negative impact on labile proteins. Visual observation indicates that operation at this agitation rate should provide adequate mixing for thawing. This conclusion is further supported by comparison of the operating level profiles for Scale 2 and 3, Figure 3.

Figure 3 shows the level (y-axis) of the ice (square), liquid (x) the total volume (solid), and contents temperature (circle) versus time on the x-axis. There are three phases of the thawing process indicated in the graphs; ascending total volume is the addition of ice without withdrawal of liquid, the flat portion is equal ice addition and liquid withdrawal, and liquid withdrawal after the conclusion of the ice addition in the descending phase. Data collection was suspended 15 minutes after no ice was present. The similarity in the Scale 2 (Figure 3, a) and Scale 3 (Figure 3, b) graphs (ratio of ice to liquid) supports the ability of the Scale 3 unit to provide heat transfer comparable to that observed in the Scale 2 unit. Another indicator of heat transfer comparability is that the product temperature during active addition of ice in both units was just above the phase transition temperature, Figure 3. This is another indication of thorough ice/liquid mixing.

Determination of the ice and liquid levels (Figure 3) was not possible by direct measurement. The complexity of the mixing made measuring the phase change from solid to liquid for a single block difficult. Colored saline-ice blocks were used to observe the movement of the ice in the vessel and to obtain an estimate of the time required for a single block to melt (at least until too small to be observed). Additionally, the surface of an ice block available for heat transfer from the thawed liquid changes over time making the determination of thawing rate for the ice very complex[3]. A mathematical model was developed to estimate the ice and liquid levels at various points in time. This was done by using the difference in the vessel jacket inlet and outlet temperatures. The temperature difference combined with the thermal properties of the heating media (heat capacity, density, and flow rate) provided a measure of the heat delivered to the system. Process development was completed with saline which provides a heat capacity close to plasma and does not take away this valuable resource from therapy production; all results were confirmed with plasma. The heat delivered over time was used in combination with the heat capacity of saline to create a mass balance showing the quantity of saline thawed; liquid and ice lines in Figure 3. The heat capacity for the phase transition of saline was determined by Differential Scanning Calorimetry (DSC) to be 271 kJ/kg. For solid ice and the thawed liquid the heat capacity of pure water was used. The aspects of the heat transfer process that could not be quantified directly (heat transfer resistance from the media to the wall, through the vessel wall, and from the wall to the vessel contents, as well as variations in heat transfer to the changing volume of the ice blocks) were lumped into a "correction factor" for consideration in a qualitative manner. To adjust the correction factor three observations were utilized; (1) visual estimates of the ice/liquid ratio at the transition from filling only to filling/withdrawing and (2) the transition at the addition of the last ice block, then (3) the time from addition of the last ice block until ice is no longer observed. The correction factor is based on Scale 2 data. The correction factor was determined by adjusting the value until the lines in Figure 3 indicated the observed ice and liquid levels at the transition points and final block addition thaw time. To provide confidence in the results this process was repeated for greater than 20 runs at Scale 2 with plasma and saline. This correction factor was then used in the assessment of Scale 3 data where the liquid/ice ratios at the indicated transition points were observed to be comparable to Scale 2, as shown in Figure 3. Additionally, the actual time required to melt the last added ice block was nearly equal, and the product temperature in the vessel for both scales remained near the phase transition temperature during ice addition. During the entire ice addition the jacket temperature was maintained at the full level (20ºC or 30ºC) for both scales. This indicates that the operating results seen for Scale 2 should be repeatable at Scale 3, including Protein Y recovery and potency.

In conclusion, the traditional method of thawing labile proteins at low agitation in a vertical vessel can be greatly improved by switching to a horizontal configuration.

A horizontal configuration utilizing multiple offset paddles has demonstrated the ability to thaw protein more quickly with less stress than a current vertical vessel. Slow mixing in the vertical unit has the unwanted effect of allowing the less dense frozen plasma to float on the surface reducing heat transfer and increasing cycle time. The horizontal vessel is able to provide interaction with the frozen plasma that gently drives the material under the liquid surface, thus increasing the opportunity for heat transfer from the jacketed wall via the liquid generated from the thawing process. Another benefit of the horizontal configuration is the thawing of a fraction of the total pool volume, and then using that working volume to feed downstream unit operations while thawing incoming frozen blocks to replenish the liquid level. This allows the time that each liter of plasma spends thawed before being separated into desired proteins to be significantly reduced as compared to a vertical vessel where a complete pool is thawed before fractionation begins. In relation to our three forces time and mechanical stress are reduced because the mix time each liter of plasma experiences is greatly reduced, while thorough mixing allows the process to run at a higher jacket temperature without concern for generation of hot spots. Addressing these design parameters provided a thawing process with demonstrated improved protein recovery for a particularly labile protein.

We would like to gratefully acknowledge Kenya Maddox (Grifols, Clayton, NC, US) and Dr. Joachim Ritter (Bayer Technology Services, Leverkusen, Germany) for their contributions to the development and understanding of the horizontal thawing process.

References:

1. Shuler M. and Kargi F. *Bioprocess Engineering: Basic Concepts* 2nd ed., Prentice Hall, Upper Saddle River, NJ, 2002
2. A. P. J. Sweere, K. Ch. A. M. Luyben and N. W. F. Kossen in *Enzyme Microbial Technology* Butterworth Publishers, London, July 1987, vol. 9, 386-398

3. Donaldson, R. D., MSc. Thesis, The University of British Columbia, 2003

(298a) Design and Characterization of Micro-Porous Hyaluronic Acid Hydrogels for Non-Viral DNA Delivery

Talar Tokatlian[1], Cynthia Cam[2], Shayne Siegman[1], Yuguo Lei[1] and Tatiana Segura[3], (1)Chemical and Biomolecular Engineering, University of California, Los Angeles, Los Angeles, CA, (2)Biomedical Engineering, University of California, Los Angeles, Los Angeles, CA, (3)Chemical and Biomolecular Engineering, University of California Los Angeles, Los Angeles, CA

Vascularization within tissue-engineered constructs remains the primary cause of construct failure following implantation. We are currently investigating two hypotheses to enhance hydrogel scaffold vascularization, both long-term mechanical support and cell-demanded release of non-viral DNA nanoparticles. Our preliminary in vivo studies show that after subcutaneous implantation for 3 weeks enzymatically degradable hyaluronic acid (HA) hydrogels have cellular infiltration only at the periphery of the hydrogel, while hydrogels with micron sized interconnected pores (micro-pore) are extensively infiltrated. Significant positive staining for endothelial markers (PECAM) was also found for micro-pore implants and not for nano-pore implants, even in the absence of pro-angiogenic factors. We believe that an open pore structure will increase the rate of vascularization through enhanced cellular infiltration and that the added delivery of DNA encoding for angiogenic growth factors will result in long lasting angiogenic signals. We have thusfar designed and characterized various pore size (30, 60, and 100 micron) micro-pore HA hydrogels loaded with a high concentration of non-viral DNA/PEI polyplexes, using a previously developed caged nanoparticle encapsulation (CnE) technique. These hydrogels allowed for long-term sustained transfection and transgene expression of incorporated mMSCs in vitro. For all investigated pore sizes, encapsulated DNA polyplexes were released steadily starting by day 4 for up to 10 days. Likewise, transgene expression was sustained over this period, although significant differences between different pore sizes were not observed. Cell viability was also shown to remain high over time, even in the presence of high concentrations of DNA polyplexes. Presently, we are using the knowledge acquired through this in vitro model to design and better predict scaffold-mediated gene delivery for local gene therapy in both subcutaneous implant and wound healing mouse models. We believe the proposed hydrogel system has applications for the controlled release of various DNA particles and other gene delivery vectors for in vivo tissue engineering and blood vessel formation.

(298b) Airway Gene Transfer and Intracellular Trafficking of Highly Compacted DNA Nanoparticles

Anthony J. Kim, Nicholas Boylan, Jung Soo Suk and Justin S. Hanes, The Center for Nanomedicine, Johns Hopkins University School of Medicine, Baltimore, MD

Despite over two decades of cystic fibrosis (CF) gene therapy research, development and clinical testing, no patient has yet been cured. Poor gene transfer to the airway epithelium has been attributed to a number of cellular barriers, including poor cellular uptake across the apical membrane, unproductive intracellular trafficking, and inefficient nuclear import. Highly compacted DNA nanoparticles, composed of single molecules of plasmid DNA compacted with block copolymers of poly-L-lysine and 10 kDa polyethylene glycol ($CK_{30}PEG_{10k}$), have shown considerable promise in human gene therapy clinical trials in the nares, but efficacy has not been established in the CF lung. Starting with the clinically tested DNA nanoparticle system, we have systematically varied the PEG MW (from 10kDa to 2kDa) and poly-L-lysine architecture and studied their effects on particle morphology, cellular uptake and intracellular trafficking, and gene transfer to lung airways in BALB/c mice. In addition, we will discuss a correlation between our intracellular trafficking data and *in vivo* gene transfer results. Together, these results represent an important step toward the rational development of an efficient gene delivery platform for the lungs based on highly compacted DNA nanoparticles.

(298c) Telecommunications Model of Lipoplex-Mediated Gene Delivery

Timothy M. Martin[1], Tadeusz Wysocki[2], Beata Wysocki[2] and Angela K. Pannier[1], (1)Biological Systems Engineering, University of Nebraska-Lincoln, Lincoln, NE, (2)Computer & Electronics Engineering, UNL, Omaha, NE

Gene delivery systems transport genetic information encoded as DNA to cells or biological systems, which then transcribe and translate that information into functional proteins. This process can be thought of as a type of nano-scale communication with the potential to directly alter endogenous gene expression and behavior with applications in functional genomics, tissue engineering, medical devices, and gene therapy. Although viral gene delivery systems are efficient at gene transfer, nonviral systems are preferred because of their low immunogenicity, inexpensive synthesis, and easy modification. In nonviral systems, the DNA is typically complexed with cationic lipids or polymers to facilitate gene transfer as the complex traverses putative intracellular barriers, the common culprits to low efficacy of these systems. Current lack of understanding of the biological mediators of gene delivery limits the design potential of new systems. Kinetic models can offer perspective and interpretation of unknown or complex cellular mechanisms involved in gene transfer. However, current models are inadequate and provide only a segmented view of gene delivery kinetics and treat complex routing as averaged or rigid events that are not biologically accurate. There is a need for new types of models that describe the complete gene delivery process with biological fitness to enable the design of more efficient systems.

In this work, we propose taking a novel approach to model the nonviral gene delivery process using telecommunications modeling. Delivery of genetic information to the cell nucleus can be considered in the same way as delivery of a packet of information to the destination computer within a packet switched computer network. In such a network, a packet containing the information is routed through several intermediate nodes before it reaches the destination, sometimes undergoing processing or transcoding within those nodes. This process is similar to gene transfer in which the complex must undergo various steps to achieve transgene expression: internalization, endosome escape, nuclear import, transcription, and translation. These steps are synonymous with intermediate nodes modeled using queuing theory with several input and output servers. Since biological processes within a cell can be considered as highly parallel, the queuing network model of gene delivery is solely dependent on kinetic constants. Through various particle tracking studies, kinetic constants are available in literature which represent the average time for complexes to travel through the various steps in the gene delivery process.

Here we used a layered communication protocol to model lipoplex delivery to an A431 epithelial cell line taking complex arrival and service time as Poisson processes. Each cell in the system was considered to consist of a queuing network as simulated lipoplexes were fed into the system and routed to servers. Stochastic averaging was used to simulate the transport process for a fixed number of individual cells. Simulation results show a high degree of agreement with experimental data, including internalization, nuclear accumulation, and transfection efficiency. Interestingly, a kinetic constant for nuclear localized plasmid transcriptional activity had to be incorporated, indicating an unaccounted for process or biological mechanism involved in gene transfer. Our novel application of queuing theory for modeling nonviral gene delivery provides insight into the pathways and mechanisms of nonviral gene transfer, so that more efficient systems can be designed.

(298d) Microrna-29b Delivery Via EpCAM Targeted Cationic Lipoplexes in Lung Cancer Treatment

Yun Wu[1], Yicheng Mao[2], Melissa Crawford[3], L. James Lee[4] and Serge P. Nana-Sinkam[3], (1)Nanoscale Science and Engineering Center, The Ohio State University, Columbus, OH, (2)Division of Pharmaceutics, College of Pharmacy, the Ohio State University, Columbus, (3)Division of Pulmonary, Allergy, Critical Care and Sleep Medicine, The Ohio State University, Columbus, OH, (4)Chemical and Biomolecular, the Ohio State University, Columbus

Lung cancer is the leading cause of cancer deaths in western countries and carries a poor overall five year survival rate. MicroRNAs are non-coding RNAs which regulate fundamental cell processes and have the potential to target multiple genes simultaneously, i.e. "one for all". MicroRNAs have been implicated in lung cancer development by serving as tumor suppressors or oncogenes. However, miRNA-based therapeutics are faced with several challenges including lack of tissue specificity, lack of optimal delivery systems, poor cellular uptake and risk of systemic toxicity. Here, we report an EpCAM (epithelial cell adhesion molecule)-targeted cationic lipoplexes based miRNA delivery system to address some of these challenges.

Among many lung cancer related miRNAs, miR-29b, a tumor suppressor, was selected as the representative therapeutic target because it has been associated with patient survival rate and sensitivity to chemotherapy. The miR-29b delivery efficiency of EpCAM targeted lipoplexes was first evaluated in A549 non-small cell lung cancer (NSCLC) cells and compared with siPORT™ NeoFX™ transfection agent (Life Technology Inc.). The expression of mature miR-29b and its 3 target genes (CDK6, DNMT3b, MCL-1) was measured using qRT-PCR and western blotting. The A549 cell growth and sensitivity to chemotherapeutic agents were characterized 24h after miR-29b overexpression. Subsequently, both the biodistribution and cytotoxicity *in vivo*were evaluated in ICR mice.

Compared to NeoFX, EpCAM targeted lipoplexes delivered miR-29b in a more efficient manner with ~10.5 fold increase in mature miR-29b expression in A549 cells. Results of flow cytometry and confocal microscopy demonstrated that the higher mature miR-29b expression was led by higher cellular uptake of EpCAM targeted lipoplexes. MiR-29b delivered by EpCAM targeted lipoplexes successfully down-regulated the expression of 3 target genes, CDK6, DNMT3b, MCL-1, by 61%, 40% and 21% respectively, but only by 20%, 22% and 17% with NeoFX. miR-29b delivered by lipoplexes significantly inhibited A549 cell growth by 62.3%, compared with 18.4% by NeoFX. Overexpression of miR-29b increased the sensitivity of A549 cells to chemotherapeutic agents. The IC50 value of cisplatin for A549 cells was 7.90uM 24h after miR-29b delivery by lipoplexes, compared to 12.94 uM by NeoFX. *In vivo*, EpCAM targeted lipoplexes achieved ~30% accumulation in lung tissue, which was ~50 fold higher than NeoFX. Lipoplexes mainly accumulated in pneumocytes, a region for lung cancer initiation. No short term cytotoxicity was observed *in vivo*. Mice systemically treated by miR-29b containing lipoplexes had a 23.6-fold induction in mature miR-29b expression in the lung compared to untreated mice.

Our results have demonstrated that the novel EpCAM targeted cationic lipoplexes may have potential as carriers for miRNA based therapeutics in NSCLC treatment. In the future, further *in vivo* miRNA-based delivery studies will be done to evaluate long-term local and systemic toxicity and target gene effects prior to clinical application.

(298e) Cellular Trafficking of Dextran Functionalized Silica Nanoparticles for Effective siRNA Delivery

Amanda P. Malefyt[1], Georgina Comiskey[2], Elizabeth Hinds[1], Stephen Lindeman[1], Gregory L. Baker[2], Christina Chan[1] and S. Patrick Walton[1], (1)Chemical Engineering and Materials Science, Michigan State University, East Lansing, MI, (2)Chemistry, Michigan State University, East Lansing, MI

Clinical applications of genetic therapies, including delivery of short, interfering RNAs (siRNAs) for RNA interference (RNAi), have been hampered by difficulties in delivering nucleic acids to cells in sufficient quantities to achieve silencing. On the cellular level, delivery efficiency is hindered by our lack of knowledge on how siRNA-vehicle complexes traffic across the plasma membrane and into the cytoplasm. Although some information is emerging, additional studies on the relationship between chemical and physical characteristics of the vehicles and their mode of uptake or trafficking within the cells are required. To address this challenge, we have developed novel dextran functionalized silica solid-core nanoparticles (NPs) with the ability to effectively deliver siRNA to human lung cancer (H1299) cells expressing an enhanced green fluorescent protein (EGFP). Through modifications to the particle size and amine content, we are able to relate the chemical and physical characteristics to changes in cellular uptake, endocytotic trafficking, and ultimately silencing efficiency. We will discuss the effect of endocytotic inhibitors on these cellular processes with respect to time and NP characteristics. Finally, we will explore the differences in silencing kinetics between lipid, polymer, and solid-core delivery vehicles.

(298f) Co-Delivery of cDNA and siRNA Using Viral/Nonviral Chimeric Nanoparticles for Synergistic Cancer Gene Therapy

Soo Kyung Cho, Chemical Engineering and Materials Science, University of California, Irvine, Irvine, CA, Shirley Wong, Pharmacology, Unversity of California, Irvine, Irvine, CA and Young Jik Kwon, Pharmaceutical Sciences, Chemical Engineering and Material Science, Biomedical Engineering, University of California, Irvine, Irvine, CA

Development of safe and efficient delivery vectors has been a pivotal demand in gene therapy. Viral vectors are popularly used in clinical trials, due to their high transduction efficiency, but their utilities have been limited by immunogenicity, pathogenicity, oncogenicity, limited gene packaging capacity, and difficulties in large-scale preparation and purification. Importantly, engineering viral vectors for versatile and multi-functional gene delivery is a cumbersome task and often infeasible. Nonviral vectors, on the other hand, offer relatively low immunogenicity, unlimited size in gene cargos, and high structural and functional flexibility for modifications, although their low transfection efficiency compared to viral counterparts is still a major drawback. Therefore, the advantages of viral vectors (i.e., high transduction efficiency) and nonviral vectors (i.e., low immunogenicity, and versatility in engineering structure and functions) were combined in viral/nonviral chimeric nanoparticles (ChNPs) with adeno-associated virus (AAV) as the viral core with acid-degradable polyketal (PK) shell as nonviral polymeric shell. The AAV core, which is one of the smallest viral vectors used in clinical gene therapy with no known linkages to human diseases, was shelled with PK shell that is designed to protect the AAV core from immune responses, promote endosomal escape, and enable the conjugation of targeting ligands on the surface. Most importantly, siRNA was incorporated in the PK shell during surface-initiated photopolymerization, in order to achieve dual modal gene delivery (i.e., simultaneous expression and silencing of multiple genes).

The core-shell structure of the resulting ChNPs, and the ordered release of siRNA and the AAV core in the presence of multiple intracellular stimuli were confirmed. Confocal microscopy demonstrated that ChNPs selectively released siRNA into the cytoplasm from the PK shell and facilitated the transnuclear localization of the AAV core. In vitro study with human lymphoma B cells showed significantly enhanced transduction by ChNPs, compared to free AAVs, with simultaneous gene silencing of additional target gene. It was also shown that the PK shell of ChNPs effectively shielded their AAV core from AAV-neutralizing antibodies and sialic acid-conjugated ChNPs selectively transduced CD22+ B-cells. Efficient and synergistic eradication of Philadelphia chromosome-positive (Ph+) leukemia was explored using ChNPs with Bim-encoding AAV core and Mcl-1 siRNA-encapsulating PK shell. Ph+ leukemia cells over-express a pro-survival Mcl-1 while down-regulating pro-apoptotic Bim. Cell apoptosis assay as well as cell growth kinetics indicated synergistic gene therapy for leukemia using ChNPs. This talk will highlight the design, synthesis, and characterization of novel gene carriers, along with a clinically relevant application.

(298g) Development of an Adenovirus Gene Therapy Vector with Improved Transduction Efficacy and Reduced Innate and Adaptive Immune Response

Adane Nigatu and Joshua Ramsey, Chemical Engineering, Oklahoma State University, Stillwater, OK

Adenoviruses-based vectors, which are currently used in more than 25 % of ongoing gene therapy clinical trials, are a promising and relatively safe class of gene therapy vectors. Adenovirus, however, has limitations that hinder its use. For example, adenovirus may initiate an innate inflammatory immune response or may be cleared rapidly in patients with preexisting immunity. Even in patients without preexisting immunity, the adaptive immune response greatly limits repeat administration of the virus. In addition, the dependency of the virus on the coxsackie-adenovirus receptor (CAR) compromises the ability of the virus to deliver genes to CAR-negative cells.

We developed a nanoparticle complex composed of adenovirus and a cell penetrating peptide (CPP)/polyethylene glycol (PEG) conjugate to reduce these drawback and produce an improved hybrid vector. The CPPs were used to enable the virus to transform cells independent of CAR, and PEG was used to prevent immune recognition, thereby reducing the innate and adaptive immune responses. The immune response towards the complexes and transduction efficiencies of the particles were optimized by varying the type of CPP (e.g., Tat, Penetratin, polyarginine and Pep1), degree of PEGylation, PEG molecular weight, and ratio of CPPs to adenovirus. In this study we found that CPP-PEG-Ad showed a 90-fold enhancement in gene transduction compared to the unmodified adenovirus when used to transduce CAR-negative cells. Moreover, the complexes were able to evade neutralizing antibodies and retained 90 % of their activity. Further, the complexes reduced the amount of proinflammatory cytokine (IL-6) produced by mouse macrophage cells. The results of the study demonstrate that modification of adenovirus with PEG and CPP can reduce immunogenic limitations associated with the virus while promoting transduction of cells not normally infected by adenovirus.

(309a) Modeling the Diffusive Behavior of 3D Stem Cell Migration

Joshua D. Cohen[1], Tyler Vlass[2], Dayong Chen[3], Ryan C. Hayward[4] and **Shelly R. Peyton**[2], (1)Chemical Engineering, MIT, Cambridge, MA, (2)Chemical Engineering, University of Massachussetts, Amherst, Amherst, MA, (3)Polymer Science and Engineering, University of Massachussetts, Amherst, Amherst, MA, (4)Polymer Science and Engineering, University of Massachusetts, Amherst, Amherst, MA

Design of 3D scaffolds that can facilitate proper survival, proliferation, and differentiation of progenitor cells is a challenge for clinical applications involving large connective tissue defects, with cell migration within scaffolds a critical process governing tissue integration. In previous work, we have quantified how physical properties of 3D scaffolds, such as pore diameter, modulus, and integrin-binding peptides, as independently tunable parameters, govern the motility and infiltration of marrow-derived stem cells (MSCs). The relationships between migration speed, displacement, and total path length were found to be strongly dependent on pore size, most likely resulting from the non-intuitive convolution of pore diameter and void chamber diameter yielding different geometric environments experienced by the cells within.

A looming challenge in the quantification of cell motility in these complex microenvironments is whether or not we will be able to *predict* the extent of cell infiltration given a set of scaffold physical parameters. We are building computational models, based on anomalous diffusion, to determine whether or not MSC migration on 2D polyacrylamide substrates, with tunable stiffness and integrin-binding peptides, is predictive of MSC migration in more complex 2.5D and 3D microenvironments. An anomalous diffusion model of cell motility exhibits superior fitting to the canonical persistent random walk (PRW) model since it is able to capture both the superdiffusive and subdiffusive behavior of MSC migration within 3D microenvironments. This model relates the mean squared displacement (MSD) to time by a power law and is characterized by two parameters, a diffusion coefficient and a scaling index. To develop a predictive quantitative model of cell motility, we are determining diffusion coefficients and scaling indices using migration paths from single cell tracking experiments and correlating these model parameters to specific physicochemical cues from the microenvironment. Experimentally, we have observed that MSC migration speed has a biphasic dependence on the density of surface-bound RGD, but only at intermediate substrate stiffness. This biphasic relationship was observed in 3D gels, but only when the pores were much larger than the average cell diameter, and therefore likely a quasi-2D environment. In smaller pore diameters, where migration was truly 3D, this biphasic dependence disappeared. We are currently designing novel "2.5D" substrates, wherein cells cannot infiltrate the scaffold, but still experience topographical cues and ridges that approximate 3D macroporous scaffolds. We hypothesize that this hybrid microenvironment will bridge our computational efforts in 2D and 3D, and allow us to predict 3D migration given the scaffolds pore size, modulus, and density of integrin-binding peptides. This information on MSC motility could be very powerful for future intelligent scaffold design for MSC-directed bone regeneration *in vivo*.

(309b) Regulatory Role of D'D3 Domain in VWF-A1 Mediated Platelet Thrombus Formation: Application towards Understanding Von Willebrand Disease

Sri R. Madabhushi[1], Kannayakanahalli Dayananda[1], Chengwei Shang[1], Jun Qu[2] and Sriram Neelamegham[1], (1)Chemical and Biological Engineering, State University of New York at Buffalo, Buffalo, NY, (2)Department of Pharmaceutical Sciences, State University of New York at Buffalo, Buffalo, NY

Introduction: von Willebrand factor (VWF) is a multimeric blood glycoprotein that plays a crucial role during blood coagulation by recruiting platelets to vascular injury site [1]. Improper regulation of VWF-platelet interaction results in various cardiovascular disorders like stroke, heart attack and other bleeding disorders called von Willebrand disease. The 3-D structural organization of VWF is believed to regulate the platelet binding function of VWF-A1 domain [2]. However, the precise mechanism by which domain level organization controls VWF function is not known. Here, we tried to address this by applying a "systems approach".

Materials and Methods: Dimeric VWF (ΔPro-VWF) was constructed by deleting the propeptide section of full-length VWF cDNA. ΔD'D3-VWF was identical to ΔPro-VWF only it lacked the VWF-D'D3 domain. Recombinant GpIbα was synthesized as a fusion protein with human IgG tail. All proteins were purified from mammalian expression systems. ΔPro-VWF was crosslinked using BS^3, a reagent that links proximal amines. This was proteolytically cleaved and subjected to high resolution tandem mass spectrometry (LTQ Orbitrap). C++ programs were written to identify cross-linked peptide pairs from the MS data. Experiments that validated the MS studies examined the effect of D'D3 domain deletion on the binding of VWF-A1 to platelet receptor GpIbα in static ELISA studies and flow chamber based shear assays. Additionally, anti-D'D3 mAbs were generated in mice and the effect of these reagents on shear induced platelet aggregation (SIPA) was studied.

Results and Discussion: Analysis of the cross-linked VWF tandem mass spectrometry data revealed three cross-linked peptide pairs linking VWF D'D3 and A1 domain. One of these A1 peptides formed a major interaction interface with platelet GpIbα. Based on this, studies were designed to assay the role of D'D3 in regulating VWF-A1 function. To this end, the deletion of the D'D3 domain (i.e. ΔD'D3-VWF) resulted in a dimeric protein that efficiently bound GpIbα even under static conditions and in the absence of ristocetin. ΔPro-VWF, on the other hand, only bound GpIbα when ristocetin was added. Under low shear conditions (1dyn/cm^2) robust platelet translocation was observed on ΔD'D3-VWF, but not ΔPro-VWF, coated surfaces. At higher shear conditions (10-20 dyn/cm^2), while platelet accumulation was comparable on the two surfaces, platelet rolling velocity on ΔD'D3-VWF was ~ 50% lower compared to ΔPro-VWF. Finally, one of the anti-D'D3 mAb generated (clone DD3.1) blocked SIPA by ~50% in the absence but not presence of ristocetin.

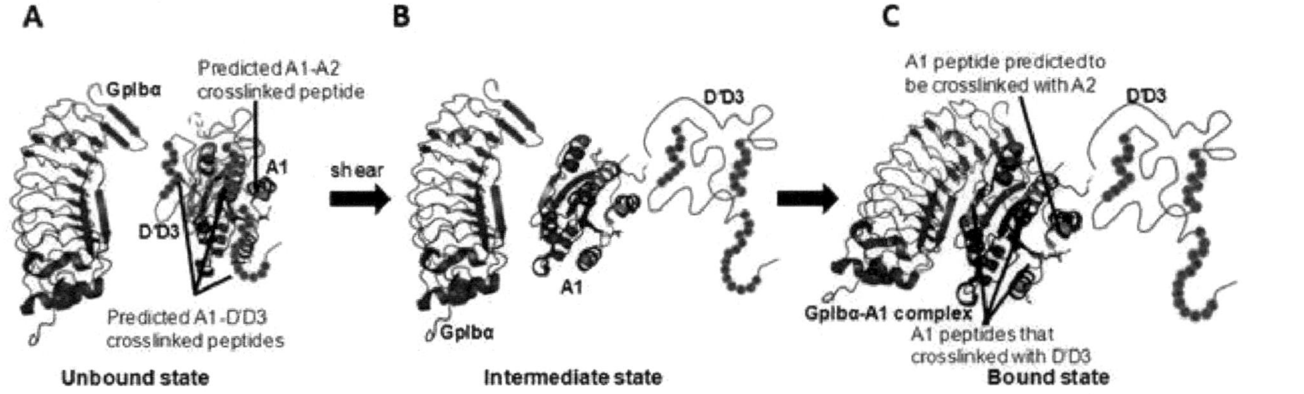

Model for VWF- GpIbα interaction under shear. A. D'D3 domain shields VWF-A1 binding to GpIbα. **B.** Upon shear, D'D3 releases A1. **C.** This enables VWF-A1 binding to GpIbα.

Conclusions: The data demonstrate that, in the native state, VWF-D'D3 domain sterically inhibits the VWF-A1 domain and this reduces the rate of VWF binding to platelets. Shear stress or agonist (e.g. ristocetin) releases this steric inhibition. This then allows VWF to bind platelet-Gp1bα efficiently. Such regulation of D'D3-A1 interaction may be a critical feature that controls platelet adhesion rates at sites of vascular injury. While von Willebrand disease (VWD) mutations have been identified, the molecular mechanism of mutation manifesting into the disease state is not well understood. Based on these results, it is likely that some of the VWD 2B or 2M mutations affecting VWF-GpIbα interaction affect the domain level regulation of VWF instead of directly affecting A1-GpIbα affinity.

References: [1] Madabhushi, S.R., et al., Blood, 2012, March 27 (*Epub ahead of print*).

[2] Singh, I et al., Biophys J., 2009;96(6):2313-20.

(309c) Thrombus Growth and Embolism On Tissue Factor-Bearing Collagen Surfaces Under Flow Role of Thrombin with and without Fibrin

Thomas Colace[1], Ryan Muthard[2] and Scott L. Diamond[1], (1)Chemical and Biomolecular Engineering, University of Pennsylvania, Philadelphia, PA, (2)Department of Chemical Engineering, Penn State University, University Park, University Park, PA

At sites of vascular injury thrombin is an important mediator in thrombus growth and stability. Using microfluidic flow devices as well as patterned surfaces of collagen and tissue factor we sought to determine the role that fibrin plays in clot stability without interfering with the production of thrombin. We deployed an 8-channel microfluidic device to study coagulation during corn trypsin inhibitor-treated (XIIa-inhibited) whole blood perfusion over lipidated tissue factor (TF) linked to a fibrillar collagen type 1 surface. Clot growth and embolization were measured at initial inlet venous (200 s^{-1}) or arterial (1000 s^{-1}) wall shear rates under constant flow rate or pressure relief mode in the presence or absence of Gly-Pro-Arg-Pro (GPRP) to block fibrin polymerization. Numerical calculations for each mode defined hemodynamic forces on the growing thrombi. In either mode at inlet venous flow, increasing amounts of TF on the surface led to a modest dose-dependent increase (up to 2-fold) in platelet deposition, but resulted in massive fibrin accumulation (> 50-fold) only when exceeding a critical TF threshold. At a venous inlet flow, GPRP led to a slight 20% increase in platelet accumulation ($p < 0.01$) in pressure relief mode with thrombi resisting ~1500 s^{-1} prior to full channel occlusion. GPRP-treated thrombi were unstable under constant flow rate, where shear forces caused embolization at a maximum shear rate of ~2300 s^{-1} (69 dynes/cm^2). In constant flow rate mode, the nonocclusive platelet-fibrin deposits (no GPRP) withstood maximum shear rates of ~29,000 s^{-1} (870 dyne/cm^2) or ~95% of full channel occlusion. For arterial inlet shear rate, embolization was marked for either mode with GPRP present when shear forces reached 87 dynes/cm^2 (~2900 s^{-1}). Under constant flow rate, platelet-fibrin deposits (no GPRP) withstood maximums of 2400 dynes/cm^2 (80,000 s^{-1}) or ~90% of full channel occlusion before embolization. Fibrin increased clot strength by 12 to 28-fold. Under pressure relief mode, ~2-fold more fibrin was produced under venous flow ($p < 0.001$). These studies define embolization criteria for clots formed with surface TF-triggered thrombin production (± fibrin) under venous and arterial flows.

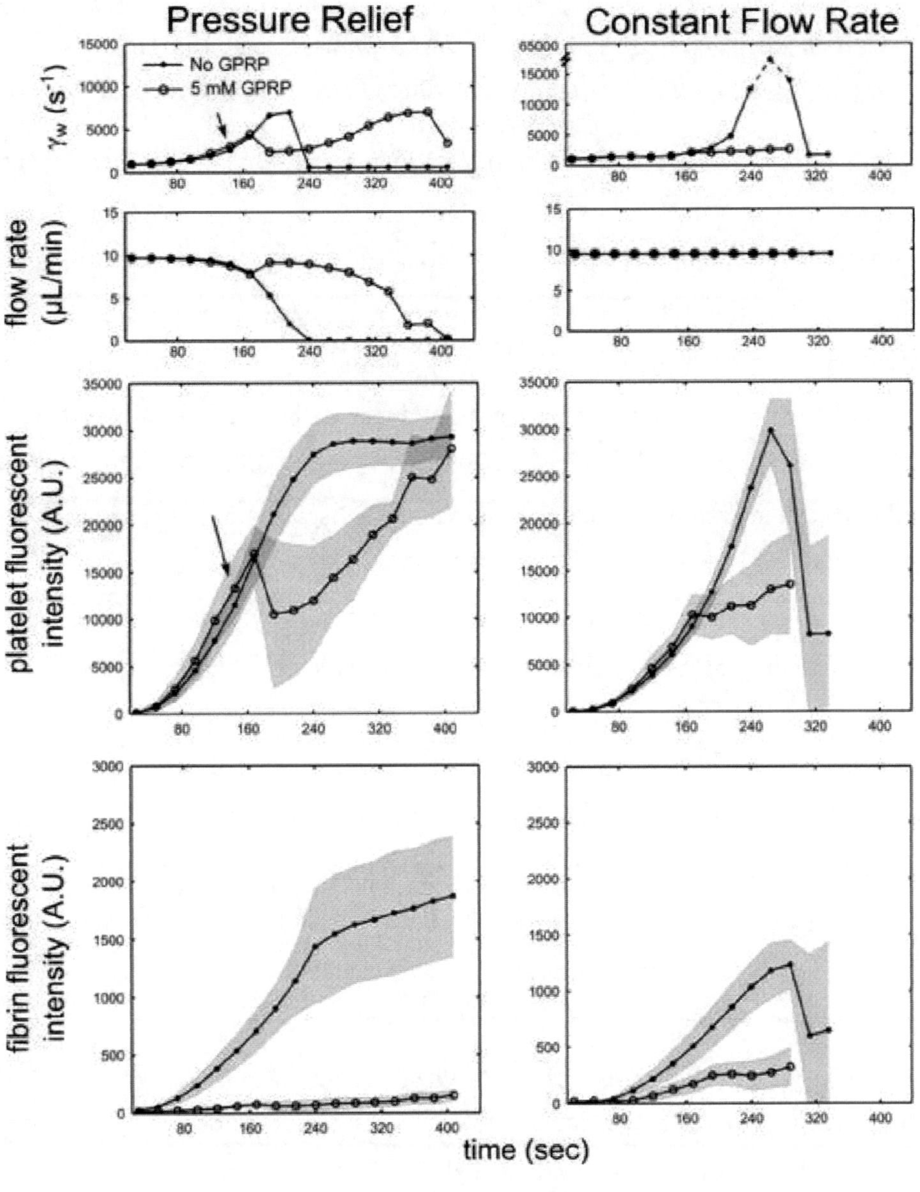

Figure 1 Whole blood (CTI 40 µg/mL) in the presence or absence of 5 mM GPRP was perfused over a fibrillar collagen surface at an initial arterial wall shear rate of 1000 s^{-1}, with (left, n=4) or without (right, n=8) pressure relief. Platelet accumulation data was used to calculate the hemodynamic conditions at the thrombus boundary based on simulation data.

A Control Engineering Perspective to Modeling Calcium Regulation and Related Pathologies

Christopher R. Christie[§], *Luke E.K Achenie*[§], *Babatunde A. Ogunnaike*[Ψ]
[§]*Department of Chemical Engineering, Virginia Polytechnic Institute and State University, Blacksburg, Virginia 24060*
[Ψ]*Department of Chemical Engineering, University of Delaware, Newark, Delaware 19716*

Calcium (Ca) homeostasis, the maintenance of a stable plasma Ca concentration in the body, is essential to normal physiological function where the total plasma Ca concentration in a healthy human body must be maintained within a very narrow range (2.2-2.4mM) [1] in the face of significant variations in Ca ingestion and/or excretion. Meeting such stringent requirements is the task of an effective, high performance physiological regulatory system that employs parathyroid hormone (PTH) and calcitriol to regulate Ca flux between the plasma and kidneys, intestines and bones. Disorders in the regulatory organs cause abnormal hormonal secretion and activity and contribute to chronic imbalances in plasma Ca levels. Furthermore, these changes in hormonal activity often lead to long-term physiological problems: for example, osteoporosis (increased loss of bone mineral density) arises as a consequence of primary hyperparathyroidism (PHPT) – a condition characterized by hyper secretion of PTH [2].

Our primary objective is to develop a quantitative understanding of how normocalcemia (the maintenance of Ca concentration in the appropriate narrow range) is achieved, and to employ such understanding to elucidate the mechanisms by which associated pathological conditions such as PHPT, hypoparathyroidism, humoral hypercalcemia of malignancy (HHM) and vitamin D deficiency affect Ca homeostasis.

In this presentation, we discuss our representation of the physiological process of Ca regulation in the form of an engineering control system block diagram with explicitly identified sensor, controller, actuators and process. Mathematical models are generated for each component sub-process by applying standard conservation of mass principles augmented with available mechanistic information. The result is a series of ordinary differential equations (ODE) that describe the dynamics of: (i) PTH in the controller; (ii) calcitriol from the kidneys in the actuator; (iii) bone cells (osteoblasts and osteoclasts) from the bone in the actuator; and (iv) plasma calcium (the "process"). The overall model is then validated with published clinical data. Pathological conditions are simulated as defects in appropriate control system components (sensor, controller and/or actuator) and the simulation results validated against published clinical data of the Ca-related disorders in question.

We demonstrate how our modeling approach provides information about important quantities that are typically not measured and are hence not available in published data sets, and how such information can be used to understand Ca homeostasis and the onset and propagation of Ca-related disorders.

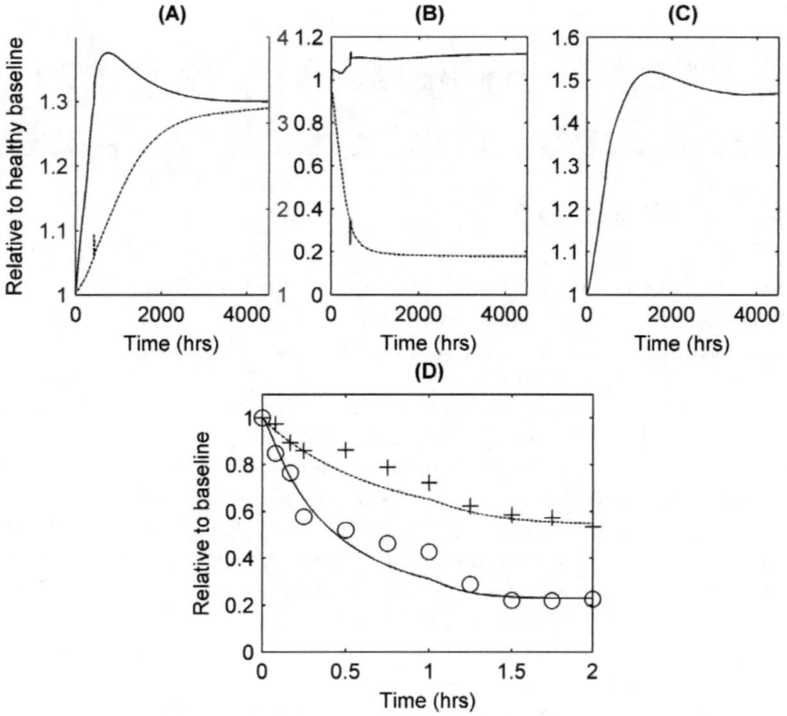

Figure 1. Controller defects: increase in PTH secretory rate (PHPT) – (A): Ca (—), PTH (--); (B) CTL (—), Cau (--); (C) bone cell ratio; (D) Comparison of clinical PTH data: Healthy: (—,o), PHPT (--- , +);

PHPT is represented in our model (Fig. 1A-C) by a defect in the controller, resulting in an increase in the PTH basal levels leading to additional increases in CTL, bone cell proliferation and reduced renal Ca excretion, consistent with clinical observations [3, 4]. This underscores the utility of our model as we are able to provide quantitative data that is usually unavailable from clinical tests. Additionally, we compare the response to hypercalcemic induction in both the healthy and PHPT models (Fig. 1D). From the modifications used to achieve the fit, we can postulate the mechanism of progression of the disease and thus provide a useful research tool in investigating, and ultimately, predicting the stage of PHPT in patients.

References

1. Raposo, J.F., L.G. Sobrinho, and H.G. Ferreira, *A minimal mathematical model of calcium homeostasis.* J Clin Endocrinol Metab, 2002. **87**(9): p. 4330-40.
2. Fraser, W.D., *Hyperparathyroidism.* Lancet, 2009. **374**(9684): p. 145-58.
3. Khosla, S., et al., *Calcium infusion suggests a "set-point" abnormality of parathyroid gland function in familial benign hypercalcemia and more complex disturbances in primary hyperparathyroidism.* J Clin Endocrinol Metab, 1993. **76**(3): p. 715-20.
4. Rubin, M.R., et al., *The natural history of primary hyperparathyroidism with or without parathyroid surgery after 15 years.* J Clin Endocrinol Metab, 2008. **93**(9): p. 3462-70.

(309e) A Quantitative Systems Approach to Identify Paracrine Mechanisms That Locally Suppress Immune Response in Melanoma

David J. Klinke, Chemical Engineering, West Virginia University, Morgantown, WV; Microbiology, Immunology and Cell Biology, West Virginia University, Morgantown, WV

Therapeutic cytokines enhance anti-tumor immunity when delivered to the tumor microenvironment. However, local immunoregulatory elements dampen the efficacy of many of these cytokines. The identity of these local mechanisms used by tumors to suppress immunosurveillance represents a key knowledge gap for improving tumor immunotherapy. From a systems perspective, local suppression of anti-tumor immunity is a closed-loop system - where system response is determined by an unknown combination of external inputs and local cellular cross-talk. Here, we hypothesized that the B16 cell line, a transplantable mouse model for spontaneous metastatic melanoma, locally suppresses the action of Interleukin-12 (IL-12) , a key cytokine that links innate to adaptive immunity. To test this hypothesis, we recreated this closed-loop system in vitro and combined quantitative high content assays, in silico model-based inference, and a proteomic workflow to identify the biochemical cues responsible for immunosuppression. Following an induction period, the B16 melanoma cell model inhibited the response of a T helper cell model to IL12. This paracrine effect was not explained by induction of apoptosis or creation of a cytokine sink, despite both mechanisms present within the co-culture assay. Tumor-derived Wnt-inducible signaling protein-1 (WISP-1) was identified to exert paracrine action on immune cells by inhibiting their response to IL12. Moreover, WISP-1 was expressed in vivo following intradermal challenge with B16F10 cells and was inferred to be expressed at the tumor periphery. This spatial pattern of WISP-1 expression is also observed in human skin melanoma tissue. Collectively, the data suggest that (1) biochemical cues associated with epithelial-to-mesenchymal transition can shape anti-tumor immunity through paracrine action and (2) remnants of the immunoselective pressure associated with evolution in cancer include both sculpting of tumor antigens and expression of proteins that proactively shape anti-tumor immunity.

(309f) A Multiscale Model of Acute Insulin Resistance in Critical Illness

Ari Pritchard-Bell, University of Pittsburgh, Pittsburgh, PA, Gilles Clermont, Critical Care Medicine, University of Pittsburgh School of Medicine, Pittsburgh, PA, Balaji Yegneswaran, Critical Care Medicine, University of Pittsburgh, Pittsburgh, PA and Robert S. Parker, Chemical and Petroleum Engineering, University of Pittsburgh, Pittsburgh, PA

Robust blood glucose regulation is an evolutionarily conserved mechanism present in many living organisms and vital to the survival of a species. These mechanisms have developed aggressive measures to prevent blood glucose from falling below threshold (hypoglycemia), while loosely regulating overshoot, which results in hyperglycemia that has potentially harmful, but less immediate, consequences. The immune response plays an important role in this regulation by overriding insulin-triggered glucose uptake to provide resources needed to fight against infection and starvation[1]. The immune activation hijacking of the body's metabolism, while evolutionarily rational for survival, induces a potentially harmful increase in glucose levels via insulin resistance. Elevated blood glucose following injury, so called "stress hyperglycemia" or "diabetes of injury", has been correlated with increased mortality rates in the ICU[2]. We aim to develop a model-based decision support system (glucose sensor, control algorithm, insulin pump, user-friendly interface) to maintain glucose homeostasis without hypoglycemia in critically ill patients. Key to the control algorithm component is a multiscale system model linking cellular response to systemic response. This does not presently exist.

We synthesize a dynamic model of glucose metabolism in relation to the acute immune response by focusing on three submodels. Our multi-scale model consists of: (i) a cellular level model of the insulin-action pathway[3]; (ii) a compartmental model describing tissue-level interaction[4]; and (iii) a novel data-driven model describing the immune response to injury. At the cellular level the antagonist(s) of insulin sensitivity deactivate the tyrosine phosphorylation cascade, following insulin binding, through serine phosphorylation[5]. We use the detailed mechanistic model proposed by Sedaghat and colleagues as a functioning insulin-responsive cell to which we add a state corresponding to the serine kinase population that acts to inhibit insulin action. The cellular model receives input from the injury model in the form of increased serine kinase levels, as well as extravascular insulin from the tissue model. The cellular insulin-responsive glucose uptake is fed back into the tissue-level model as a variable glucose clearance rate. The tissue-level model from Roy[4]also accounts for exercise and fatty acids which can be used as inputs for the injury model.

The injury model characterizes the immune response to trauma using biometrics, cytokines and stress-hormones as inputs that are believed to influence serine kinase activity. Data from APACHE III scores (a recognized metric describing severity of illness[6]) are used as inputs that have been correlated to insulin sensitivity[7]. Fatty acids from the tissue model are classified by their inflammatory characteristics and accordingly propagate the immune response. Stress-induced hormones such as catecholamines and glucocorticoids[8]also influence insulin resistance, and these are described in the injury model as additional inputs to the cellular model to attenuate insulin-dependent glucose uptake.

The wide range of patient conditions and biomolecules contributing to stress hyperglycemia makes accurate characterization a serious challenge. The composite model allows us to depict the overall dynamics of glucose metabolism while detailing important molecular events. Analysis at the cellular level allows us to identify dynamic micro-scale factors influencing biological regulation of systemic glucose and insulin in patients recovering from trauma. Using multiple scales we are able to describe key phenomena relevant to glucose control, as applied to critical care patients. Molecular-level insight into stress hyperglycemia could reveal key aspects of patient specific acute insulin resistance which could be used in a model-based decision support system for controlling glucose in the intensive care unit.

1. Fernandez-Real, J.-M., et al. (1999). Insulin Resistance and Inflammation in an Evolutionary Perspective: the Contribution of Cytokine Genotype/Phenotype to Thriftiness. *Diabetologia,*42: 1367-1374

2. Van den Berghe, G. et al., (2001). Intensive Insulin Therapy in Critically Ill Patients. *The New England Journal of Medicine*. 345:1359-1367.

3. Sedaghat, A. R. et al. (2002). A Mathematical Model of Metabolic Insulin Signaling Pathways. *American Journal of Physiology Endocrinology and Metabolism*, 283: E1084-E11101

4. Roy, Anirban (2008). Dynamic Modeling of Free Fatty Acid, Glucose, and Insulin During Rest and Exercise in Insulin Dependent Diabetes Mellitus Patients. Ph.D dissertation, University of Pittsburgh.

5. Kitano, H. et al. (2004). Metabolic Syndrome and Robustness Tradeoffs. *Diabetes,*53 Suppl. 3: S6-S15.

6. Knaus, W. et al., (1991). The APACHE III Prognostic System. *CHEST*. 100(6): 1619-1636.

7. Zauner, A. et al. (2007). Severity of Insulin Resistance in Critically Ill Medical Patients. *Metabolism.* 56(1): 1-5.

8. Li, L. et al. (2009). Acute Insulin Resistance Following Injury. *Trends in Endocrinology and Metabolism.* 20(9): 429-435.

(309g) Systems Biology of Host-Pathogen Interactions

Anu Raghunathan, Chemical Engineering, National Chemical Laboratory, Pune, India, Sookil Shin, Mount Sinai School of Medicine, NY, NY and Simon Daefler, Division of Infectious Diseases, Mt Sinai School of Medicine, New York, NY

Francisella tularensis is an ill-characterized pathogen and a potential biowarfare agent necessitating a better understanding of its metabolism and virulence mechanisms. Genome scale constraints based metabolic models provide a systems framework to study metabolism, infection and pathogenesis. We have developed a metabolic network reconstruction for *F. tularensis* using the current genome sequence annotation and biochemical legacy data. Several mathematical techniques like Flux balance analysis (FBA), Flux Variability Analysis (FVA), Robustness Analysis (RA) were used to explore the metabolic capabilities of the *in silico* pathogen under *in vitro* and *in vivo* conditions. Gene expression, growth and physiological data were obtained experimentally and used as additional constraints in the model to compute and predict functional states of the pathogen. Integrating this data into the model as constraints allow us to reduce the solution space representing cell behavior and predict cellular function with more accuracy. *Francisella* was capable of growth with different specific growth rates on glucose and alternate carbon sources including glycerol, xylose, ribose, fructose and arabinose in a chemically defined medium with differential and preferential consumption of amino acids thus validating our model prediction that they can make up the bulk of the biomass. Gene expression of selected metabolic genes at specific time points (12hr and 24hr pi) of *francisella* during infection provide a snapshot and implicate certain metabolic pathways as critical to growth and survival in the host cell. Several gene transcripts detected in this data set were present in the optimal reactome calculated using FVA. Such an analysis provides insights into the lifestyle of the bacterium and the metabolic dependence on the host. Iterative model building can further help refine our understanding of the intracellular milieu encountered by the bacterium inside the host-cell and point towards genes essential in growth, survival and pathogenesis. Collectively, the results presented herein suggest an effective systems biology strategy of combining *in silico* modeling with experimental technologies to enhance biological discovery and discover novel drug targets.

(309h) Modulating Lipid Fate Controls Lipotoxicity in Palmitate-Treated Hepatic Cells

Alexandra K. Leamy, Chemical & Biomolecular Engineering, Vanderbilt University, Nashville, TN and Jamey D. Young, Chemical and Biomolecular Engineering, Vanderbilt University, Nashville, TN

As the American waistline continues to expand so too does the burgeoning health crisis associated with obesity, diabetes and the all-encompassing metabolic syndrome. The pathogenesis of these now common conditions involves abnormal accumulation of lipids (steatosis) in non-adipose tissues due to elevated free fatty acid export from adipose tissue. In the liver, chronic steatosis and saturated fatty acids (SFA) overexposure can result in dysregulation of ER function and oxidative stress. However, the lipid metabolic pathways and mechanisms responsible for these deleterious alterations remain largely undefined. Our central hypothesis is that overexposure to SFAs (e.g., palmitate) leads to increased saturation of ER membrane phospholipids, which alters membrane stability and induces ER stress.

In order to better characterize the phenotypic response of hepatic cells to SFA overexposure, we have sought to systematically perturb the three main pathways of fatty acid utilization and disposal: (1) phospholipid (PL) synthesis and remodeling, (2) energy generation by beta-oxidation, and (3) triglyceride (TG) synthesis. Experimentally, we simulated hepatic lipotoxicity using palmitate-treated H4IIEC3 rat hepatocytes as an *in vitro*model. This system demonstrated significantly higher levels of reactive oxygen species (ROS) accumulation, endoplasmic reticulum (ER) stress associated protein release (CHOP), and cell death as well as increased saturation of membrane phospholipid species. Using treatments designed to modulate specific lipid metabolic pathways, we have found that blocking phospholipid breakdown and turnover (using phospholipase inhibitors) or enhancing fatty acid disposal (using AICAR to stimulate beta-oxidation) can suppress the lipotoxic effects of palmitate.

A detailed investigation is currently underway to elucidate the mechanisms by which these treatments are able to reverse the lipotoxic effects of SFA overexposure. We are applying isotopically labeled ^{13}C-palmitate to trace the fate of exogenous fatty acids using metabolic flux analysis (MFA) and lipidomic profiling. This will enable us to identify pathways that are responsible for the observed metabolic phenotypes and develop a better understanding of the mechanisms leading to ROS accumulation and lipoapotosis in the presence of elevated SFAs. This understanding is expected to provide more effective and targeted therapies for treatment of fatty liver disease and related complications associated with diabetes and obesity.

(312a) Analytical Model of Local Distribution of Chemicals in Tissues with First Order Rate Metabolism Kinetics

Alexander Golberg, UC Berkeley, Berkeley, CA; CEM/MGH/Harvard Medical School, Boston, MA

Analytical model of local distribution of chemicals in tissues with first order rate metabolism kinetics.

Alexander Golberg,

Department of Mechanical Engineering, *Etcheverry* Hall, 6124, University of California at Berkeley, Berkeley, CA 94720 USA

Email: agolberg@gmail.com

Tel: +1 -510-2216557

Abstract

The profile of regional distribution of chemicals in tissue is important for the fundamental analyses of metabolism, morphogenetic growth control and the development of novel regional therapies. One of the challenges in tissue and whole organ research is predicting local chemical distribution. Here we introduce an analytical approach for the temporal and spatial regional distribution mapping of chemicals in metabolically active, perfused tissues. The new aspect of this model is in local chemical concentration dependence on of tissue perfusion rate, in addition to diffusion and metabolism rates. Using Duhamel theorem we performed dimensionless and dimensional analysis for local chemical distribution and report on steady-state and transient solutions for tissues with the first rate kinetic order of clearance by perfusion and metabolism. The predictions of our model are in good correlation with clinically observed data on percutanious ethanol ablation volumes of liver. Our study shows the importance to incorporate tissue perfusion rate to the fundamental understanding of *in vivo* intercellular signaling and treatment planning of regional therapies, which include local drug injection, which include electrochemotherapy, percutanious ethanol ablation and regional anesthesia.

Table 1. Model inputs and values investigated in the parametric study.

Variable	Physical Property	Units	Investigated range
t	Time from the compound introduction	s	1E-3-1E+3
x	Distance from the source	m	1E-3-1E-1
B	Boundary concentration	M	1

D	Diffusion coefficient of a chemical compound in the tissue	$m^2 \cdot s^{-1}$	5E-6
w_p+k_b	Average tissue perfusion rate and first order metabolic rate.	s^{-1}	1E-4 – 7E-2
ε	Minimal fraction of the chemical compound that causes to a desired effect		4E-1-9.5E-1

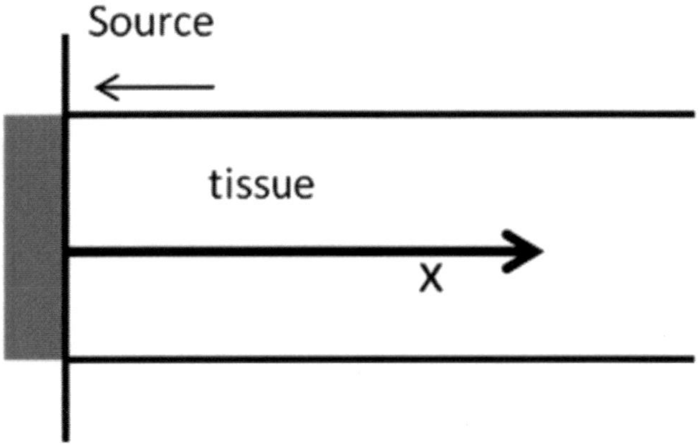

Fig.1 Analyzed physical system.

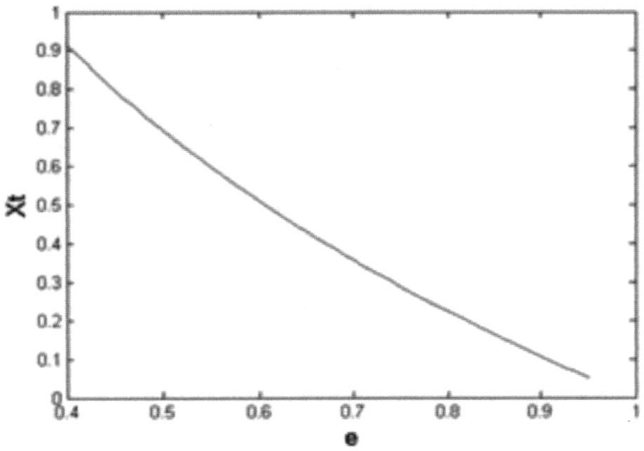

a

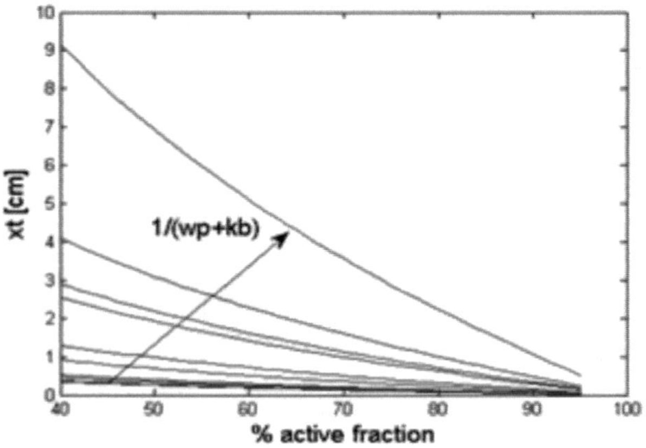

b

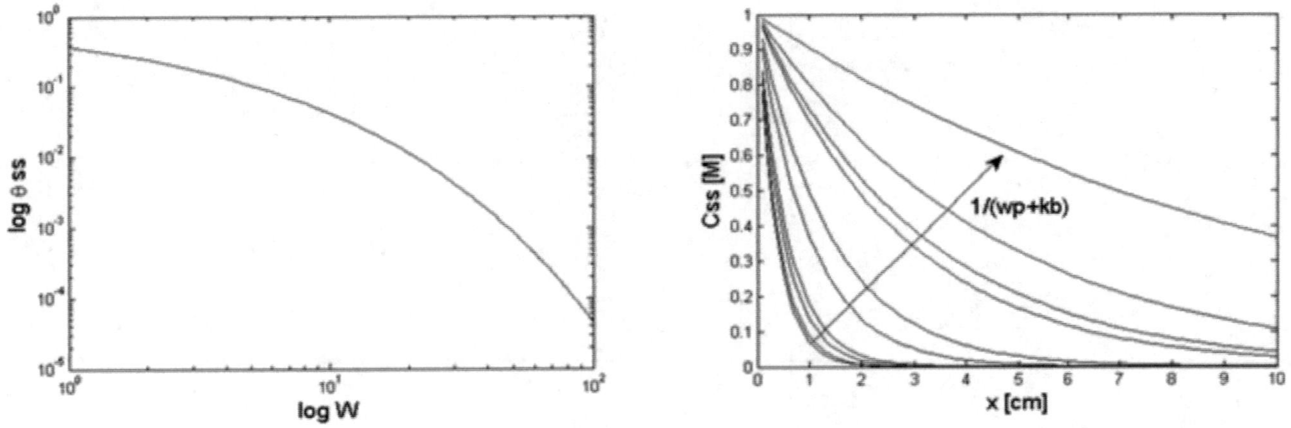

c d

Fig.2 Parametric study results. **a.** Dimensionless drug penetration depth (X_t) as a function of minimum active concentration(ε). **b.** Drug penetration depth as a function of a drug fraction in %. **c.** Dimensionless steady state concentration as a function of dimensionless metabolic rate (W). **d.** Drug steady-state concentration (C_{ss}) for various perfusion and absorption rates.

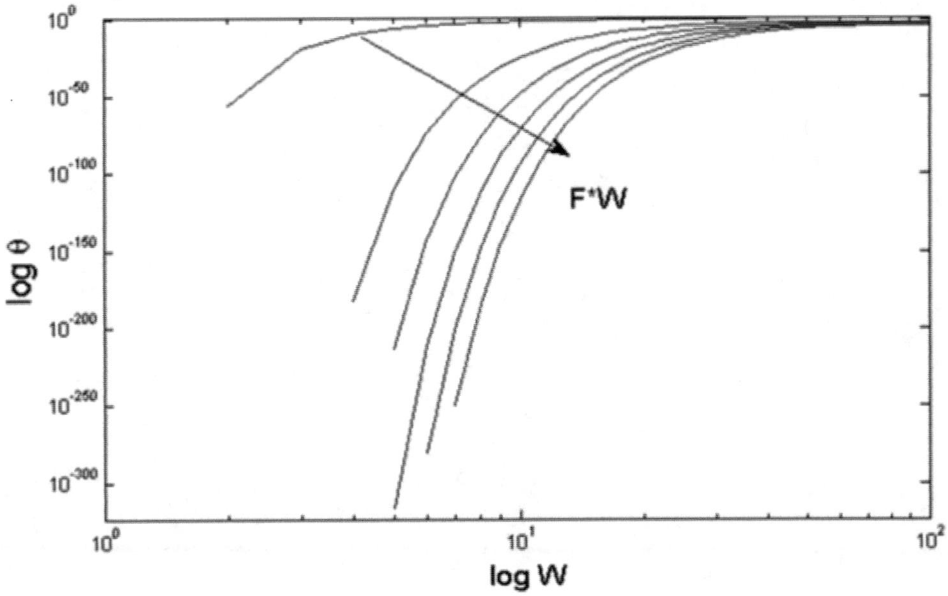

Fig.3 Parametric study results. Transient dimensionless concentration as a function of dimensionless perfusion rate (W) and time (F).

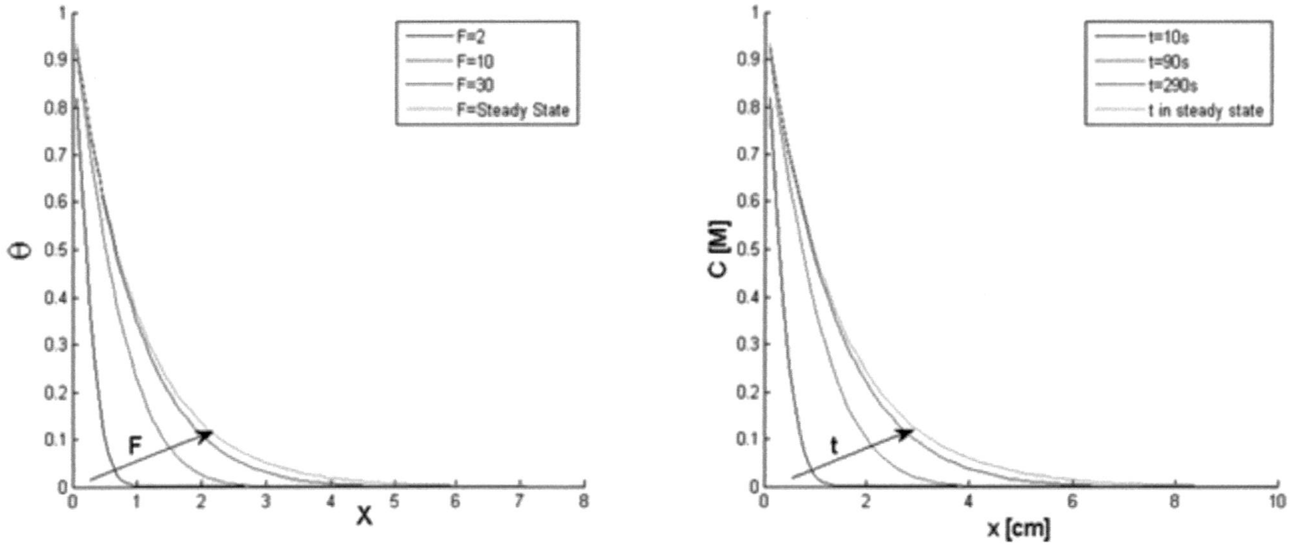

a b

Fig.4 Parametric study results. a. Transient dimensionless concentration as a function of dimensionless time and location for dimensionless perfusion rate *W*=50 **b.** Transient concentration as a function of time and location for perfusion rate $w_p+k_b=5*10^{-3}$ s^{-1}.

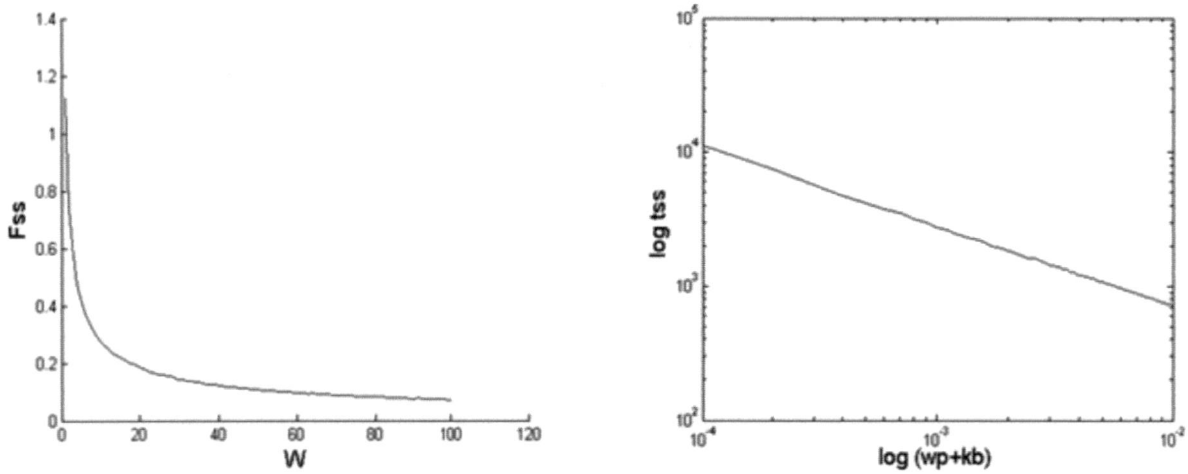

a b

Fig.5 Parametric study results. a. Dimensionless time to steady-state (F_{ss}) as a function of dimensionless metabolic rate (W) **b**. Time to reach steady state (t_{ss}) as function of perfusion (w_p) and metabolism rates (k_b).

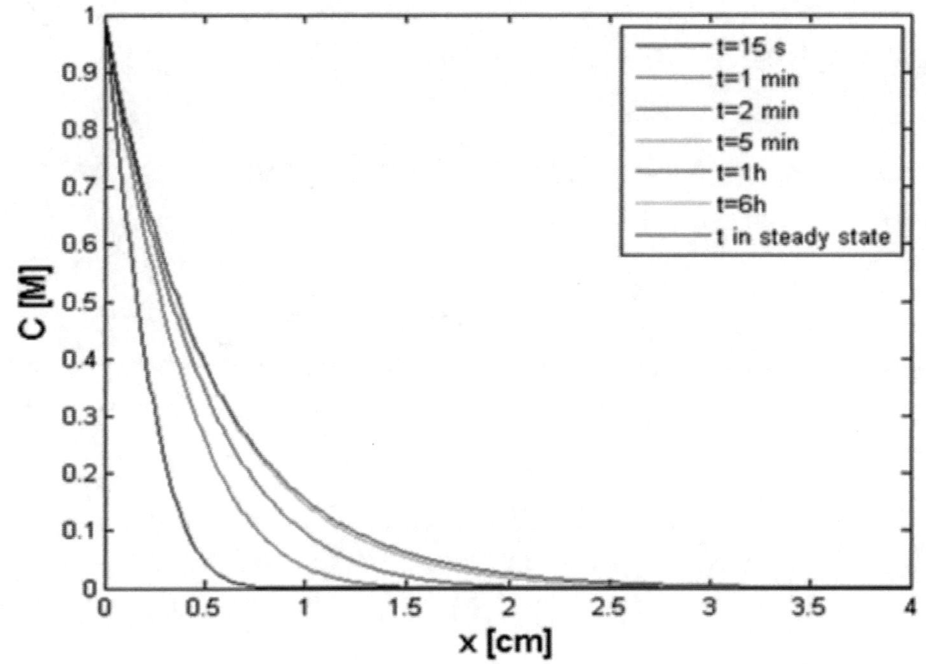

Fig.6 Practical example. Ethanol concentration in a liver as a function of location and time during percutaneous ethanol injection therapy. The system parameters appear in Table 2. 98% ethanol was injected in the liver (Boundary condition).

Table 2. Diffusion and metabolic parameters for percutaneous ethanol injection of liver model.

Variable	Physical meaning	Values
D	Diffusivity	2.2E-3 cm s^{-1}
k_b	First order kinetics of ethanol absorption in liver	8.1E-4 s^{-1}
w_p	Average perfusion rate of small rat liver tumor	45.5 ml 100$gr^{-1}min^{-1}$

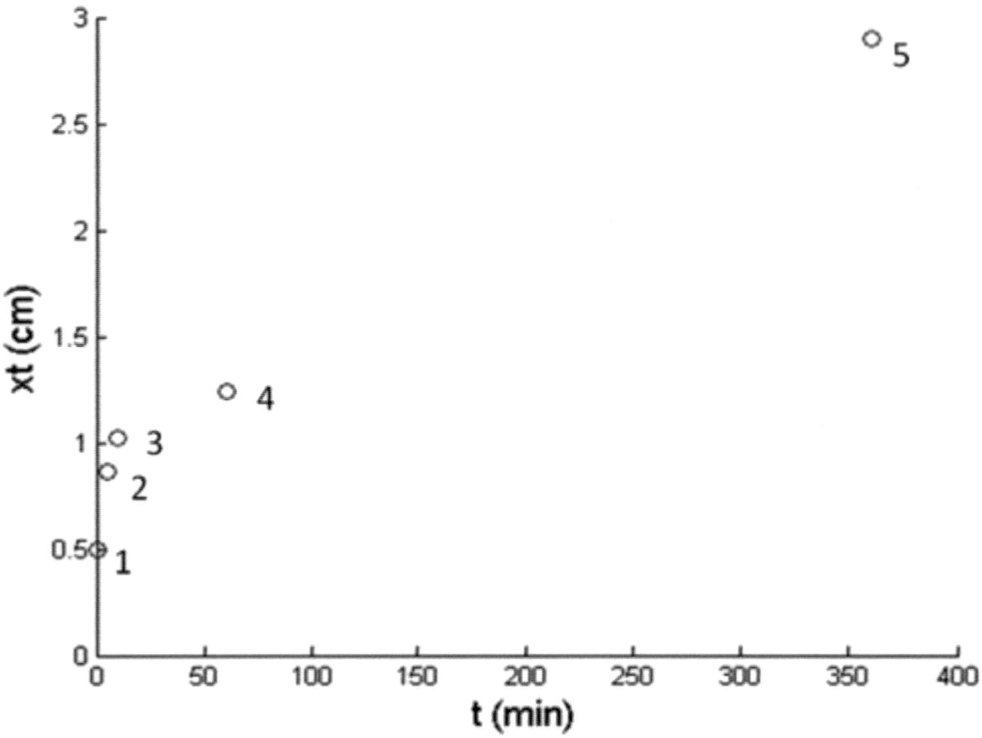

Fig.7 Practical example. Calculated distance from the injection point at which ethanol causes liver tumor ablation by necroses (points 1-4) and apoptosis (point 5) as described in Table 3:

Table 3. Toxic ethanol concentration for various exposure durations.

Time [s]	Critical concentration %	ε	Point N on Fig. 6
15	40	0.4	1
300	20	0.2	2
600	15	0.15	3
3600	10	0.1	4
21600	0.046	0.0046	5

(312b) Modeling the Superovulation Stage in in-Vitro Fertilization (IVF)

Kirti Maheshkumar Yenkie, Bioengineering, University of Illinois, Chicago, Chicago, IL, Urmila Diwekar, Vishwamitra Research Institute, Center for Uncertain Systems: Tools for Optimization and Management, Clarendon Hills, IL and Vibha Bhalerao, Jijamata Hospital, Nanded, India, Nanded, India

Abstract

In-vitro fertilization (IVF) is the most common technique in assisted reproductive technology and in most cases the last resort for infertility treatment. It has four basic stages: superovulation, egg retrieval, insemination/fertilization and embryo transfer. Superovulation is a drug induced method to enable multiple ovulation per menstrual cycle. The success of IVF majorly depends upon successful superovulation, defined by the number and similar quality of eggs retrieved in a cycle.

The model is adapted from the theory of batch crystallization. The aim of crystallization is to get maximum crystals of similar size and purity, while superovulation aims at eggs of similar quality and size. The rate of crystallization and superovulation are both dependent on the process conditions and varies with time. Thus, model formulation for multiple ovulation is on parallel lines to crystal formation in a batch process and is modeled as such. The kinetics of follicle growth is modeled as a function of injected hormones and the follicle properties are represented in terms of the moments. Hence, modeling of this stage in terms of distribution of eggs (oocytes) obtained per cycle involving the chemical interactions of drugs used and the conditions imposed on the patient during the process provides a basis for predicting the possible outcome. The results from the model prediction were verified with the known data from Jijamata Hospital Nanded, India. The predictions were found to be in well matched with the actual observations.

Thus, a phenomenon currently based on trial and error will get a strong base in terms of a predictive model. It will help the patient to decide whether to undergo superovulation or start the IVF from donor eggs, which in turn would save the patient from financial loss as well as emotional distress.

1. Introduction

Around 80 million people in the world are suffering from infertility issues. The rate of fertility is constantly declining in the developed nations due to late marriages, postponed childbearing and primary infertility. On the contrary, in the developing world the reasons for infertility involve prevalence of sexually transmitted diseases, infections which increase the rate of secondary infertility. Childlessness is often stigmatized and leads to profound social suffering for women in the developing nations (de Melo Martin, 1998).

1.1. In-vitro fertilization:

It is a process by which oocytes or egg cells are fertilized by a sperm outside the body in a laboratory simulating similar conditions in the body and then the fertilized eggs are implanted back in the uterus for full term completion of pregnancy.

Four stages in IVF:

i. Superovulation: It is method to retrieve multiple eggs using drug induced stimulation. In normal female body only one egg is ovulated per menstrual cycle, but with the use of fertility drugs and hormones, more number of eggs can be ovulated per cycle.

ii. Egg collection (retrieval): On the maturation of the multiple eggs produced in the previous stage, the eggs are retrieved through special techniques like ultrasonically guided transvaginal oocyte retrieval.

iii. In vitro fertilization (inseminaton/fertilization): Fertilization is done in the incubator using the retrieved oocytes and sperms. The conditions are maintained so as to mimic the invivo environment.

iv. Embryo transfer: The fertilized embryos are implanted into the uterus via a non-surgical technique using ultrasound guidance.

The major cost of IVF is associated with the superovulation stage where expensive drugs are used and almost daily monitoring is required. Success of this stage in terms of number and quality of eggs affects the outcome of IVF. In this work, we concentrated on modeling this stage and the approach is presented in the next section.

1.2. Superovulation:

In this work, we follow the analogy between batch crystallization and superovulation to model the process.

Analogy between Superovulation and Batch crystallization:

The moment model for follicle number and size is adapted from the concept of batch crystallization (Q. Hu. et. al., 2005) based on the analogy between batch crystallization and superovulation presented in Table 1.

Table 1: Analogy between batch crystallization and IVF superovulation stage

Batch Crystallization	Superovulation (IVF stage I)
Production of multiple crystals	Production of multiple oocytes or eggs
Crystal quality is determined in terms of size distribution and purity	Oocyte quality is determined in terms of no abnormalities, similar size.
The rate of crystallization or crystal growth varies with time and process conditions	The rate of ovulation or oocyte growth varies with time and drug interactions
Process is affected by external variables like agitation, and process operating variables like temperature, pressure, etc.	Process is affected by externally administered drugs and body conditions of the patient undergoing the process

The superovulation follicle growth model in general resembles greatly to the growth of seeded batch crystals. Thus, the moment model for both the processes remain the same the growth term which is a function of process variables like temperature and supersaturation in batch seeded crystallization will become a function of medicinal dosage in case of superovulation process.

2. Model details

Due to ovarian stimulation using externally injected hormones the number of follicles activated to enter into the ovulation stage are more in number as compared to a single follicle in a normal menstrual cycle. From the current data on successful superovulation for patient 1, organized in Table 2; it can be observed that during FSH dosage regime, as the time progresses the size of the eggs increase.

Table 2: Variation of Follicle size (diameter) with time and FSH dose

Size range (mm) dia Sr. No	↓ → Days	Day 1	Day 2	Day 5	Day 7	Day 9
1.	0-4	15	8	6	0	0
2.	4-8	3	6	4	2	0
3.	8-12	0	4	10	14	4
4.	12-16	0	0	0	2	11
5.	16-20	0	0	0	0	3
6.	20-24	0	0	0	0	0
	FSH dose	0	300	300	300	225

2.1. Model Assumptions:

The rate expression for follicle growth is dependent on FSH administered. Thus, we can write the growth term as;

$G=kCfsh\alpha$ $G = kC_{fsh}^{\alpha}$ (1)

Assuming moment model, we consider only the first seven moments; zeroth moment corresponding to follicle number, the first moment corresponding to follicle size and the other 5 moments. 1st to 6th moments are being used since they help in recovering the size distributions more precisely as against lower number of moments.

2.2. Model equations (Q. Hu. et. al., 2005):

$\mu0=constant$ $\mu_0 = constant$ (2)

$d\mu1dt=Gt\mu0t$ $\frac{d\mu_1}{dt} = G(t)\mu_0(t)$ (3)

$d\mu2dt=2Gt\mu1t$ $\frac{d\mu_2}{dt} = 2G(t)\mu_1(t)$ (4)

$d\mu3dt=3Gt\mu2t$ $\frac{d\mu_3}{dt} = 3G(t)\mu_2(t)$ (5)

$d\mu 4dt=4Gt\mu 3t$ $\frac{d\mu_4}{dt} = 4G(t)\mu_3(t)$ (6)

$d\mu 5dt=5Gt\mu 4t$ $\frac{d\mu_5}{dt} = 5G(t)\mu_4(t)$ (7)

$d\mu 6dt=6Gt\mu 5t$ $\frac{d\mu_6}{dt} = 6G(t)\mu_5(t)$ (8)

Conversion of the data available on follicle number and size to moment using the expression given in literature by Flood, 2002:

$\mu i=nir,trinDri$ $\mu_i = \sum n_i(r,t) r_i^n \Delta r_i$ (9)

Here, μ_i = i^{th} moment

$n_i(r,t)$ = number of follicles in bin of mean radius 'r' at time 't'.

r_i = mean radius of ith bin

Δr = range of radii variation in each bin

3. Results:

We integrate the equations 2-8 for predicting the kinetic constants in the follicle growth expression. Later we use non-linear optimization algorithm to predict the values of these kinetic constants along with the integration constants obtained after integrating the set of moment equations. In real practice, the model will be calibrated with the first two days of data and then used for prediction of the complete cycle.

3.1. Model Validation:

The current moment model predicts the moment values, however our final output desired is the follicle size distribution, thus in model validation the approach to obtain follicle size distribution from moment values are shown. The method is adapted from the literature by Flood 2002; where he shows the method to recover particle size distribution from moments in batch crystallization. Using the model predicted moment values we evaluate n(r,t) and compare with the actual data to check the model accuracy. We plot the follicle size distribution for four patients for various days. The experimental size distribution is shown by symbols while the continuous curve shows the model predicted values after using the inversion method.

4. Conclusion:

The moment model developed for IVF superovulation predicts the follicle size distribution which is in well agreement with the actual size distribution seen in the IVF cycle data. The model can be used to predict the outcome. This will reduce the almost daily requirement of testing. The model can also provide a basis for predicting the optimum dosage for the desired outcome from the superovulation stage. The model used here is a very basic model and the complexities present in the patient are not considered. Later, we aim to include these complexities and model the system uncertainties, using more data for analysis, modeling and validation.

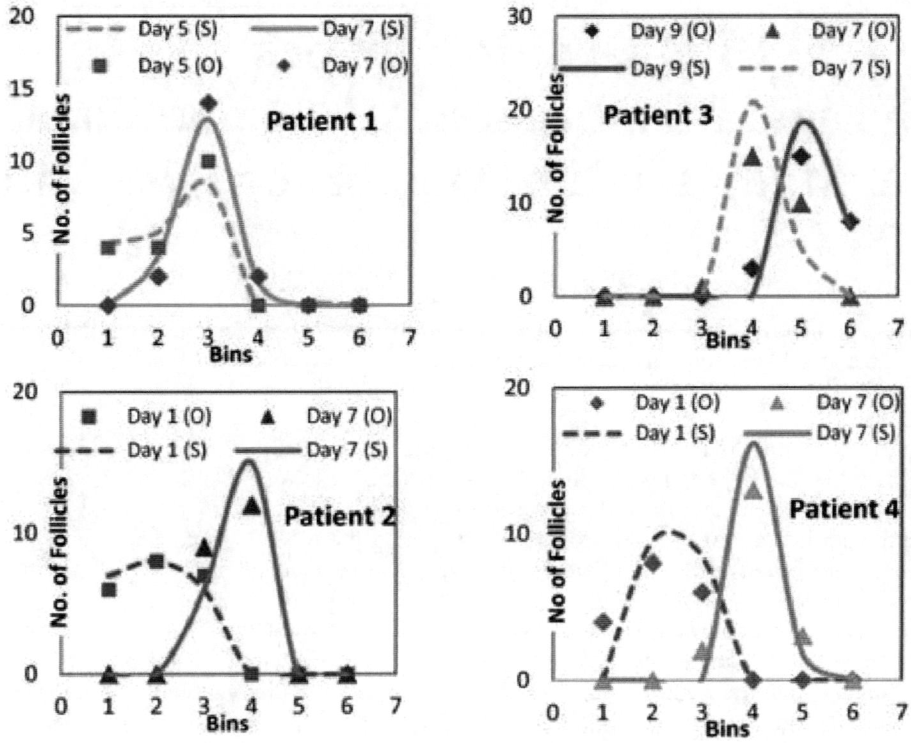

Fig 1. Follicle size distribution for four patients

References

de Melo-Martin, I., 1998. "Ethics and Uncertainty: In Vitro Fertilization and Risks to Women's Health" Risk: Health, Safety & Environment 201.

Flood, A. E., 2002. "Thoughts on recovering particle size distributions from the moment form of the population balance." *Dev. Chem. Eng. Mineral Process,* Vol. 10, No. 5/6, pp. 501-519.

Q. Hu, S. Rohani, A. Jutan, 2005. “Modelling and optimization of seeded batch crystallizers” *Computers and Chemical Engg.* Vol. 29, pp. 911-918.

(312c) Interactions of PEO-PPO-PEO Block Copolymers with Lipid Membranes: A Computational and Experimental Study Linking Membrane Lysis with Polymer Structure

Paola Carbone, School of Chemical Engineering and Analytical Science, The University of Manchester, Manchester, United Kingdom and Selina Nawaz, Chemical Engineering and Analytical Science, University of Manchester, Manchester, United Kingdom

Poly(ethylene oxide) (PEO), and poly(propylene oxide) (PPO) block co-polymers (PEO-PPO-PEO, sold as Pluronics, Poloxamers, Tetronics) are a widely-used class of amphiphilic materials for different biological applications. In fact for certain members of the Pluronics series, the interactions of block segments with living cells alter the lipid membrane properties and facilitate the permeation of drugs. A fuller understanding the molecular mechanisms underpinning these interactions is essential for ensuring their safety and efficacy in biomedical applications and to inform the design of new amphiphilic copolymers for potential use in a clinical setting. In this paper, by means of atomistic molecular dynamics simulations and membrane lysis assays, we investigate the relationship between the molecular conformations of a subset of the Pluronic copolymers (L31, L61, L62 and L64) and their haemolytic activity. Our computational studies suggest that the hydrophilic blocks in these copolymers interact with the polar head groups of lipid molecules, resulting in a predicted modification of the structure of the membranes. Parallel membrane lysis assays in human erythrocytes indicate differences in the rates of haemolysis, as a result of incubation with these polymers, that correlate well with the predicted interactions from the atomistic simulations. The computational data thus provide a putative mechanism to rationalize the available experimental data on membrane lysis by these copolymers and quantitatively agree with haemoglobin release endpoints measured when copolymers with the same molecular weight and structure of those modelled are incubated with erythyocytes. The data further suggest some new structure-function relationships at the nanoscale that are likely to be of importance in determining the biological activity of these otherwise inert copolymers.

(312d) Computational Study of Drug Transport in Realistic Models of Solid Tumour

Wenbo Zhan[1], Wladyslaw Gedroyc[2] and Xiao Yun Xu[1], (1)Department of Chemical Engineering and Chemical Technology, Imperial College London, London, United Kingdom, (2)Department of Medicine, Imperial College London, London, United Kingdom

Background: The efficacy of anticancer therapy is often offset by serious side effects. This is largely due to toxicity of the chemotherapy drug and its undesirable uptake in *vivo*. Various attempts have been made to retain the anticancer effect while reducing toxicities, including continuous infusion and liposomal delivery of chemotherapy drugs. Since the outcome of an anticancer therapy depends strongly on the pharmacokinetics of the drug as well as the properties of tumour and its surrounding tissue, rational criteria for choosing an optimal delivery mode and dose for different types of tumours are not yet available.

Methods: In this study, a computational model is developed which incorporates real tumour geometry reconstructed from magnetic resonance (MR) images, and the key physical and biochemical processes involved. These include time-dependent plasma clearance, drug transport through the blood and lymphatic vessels, extracellular drug transport (convection and diffusion), binding with proteins, lymphatic drainage, interactions with the surrounding normal tissue and drug uptake by tumour cells. Anticancer efficacy is evaluated based on the faction of survival tumour cells by directly solving the pharmacodynamics equation using the predicted intracellular doxorubicin concentration. Comparisons are made of the predicted efficacies of different modes of drug administration and doses in a clinical relevant range. As a typical anti-tumour drug, doxorubicin is employed in the computational model.

Results and Conclusion: Our results show that liposome-mediated drug delivery performs best in reducing drug concentration in normal tissues, which will help lower the risks of associated side effects. Bolus injection leads to the highest extracellular concentration in both tumour and normal tissues, giving the worst efficacy. Continuous infusion can effectively reduce the extracellular concentration while increasing the intracellular concentration (shown in Fig. 1), which mau help reduce the side effects caused by high concentration in normal tissues. As shown in Fig. 2, rapid continuous infusions lead to quick cell killing in the early phase of drug administration, but slow infusion can kill more tumour cells over a longer time period. Moreover, continuous infusion over 2-hour duration seems to offer the highest intracellular concentration. In addition, the effects of tumour size and shape have been investigated based on patient-specific MR images.

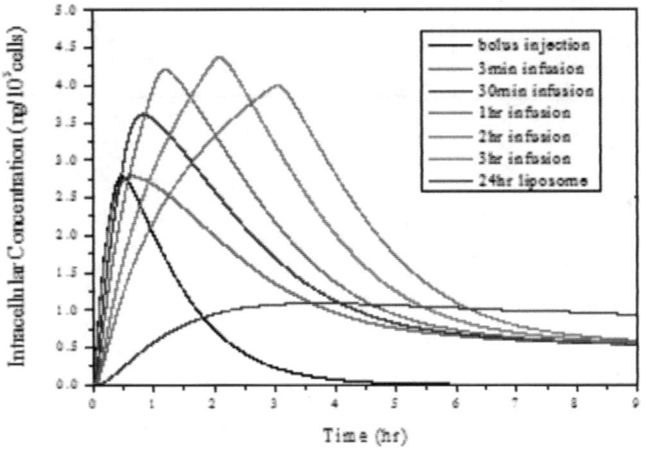

Fig. 1 Comparison of intracellular concentration among different modes of drug administration (bolus injection, continuous infusion over durations of 3 minutes, 30 minutes, 1 hour, 2 hours, and 3 hours, respectively, and liposomal release over 24 hours)

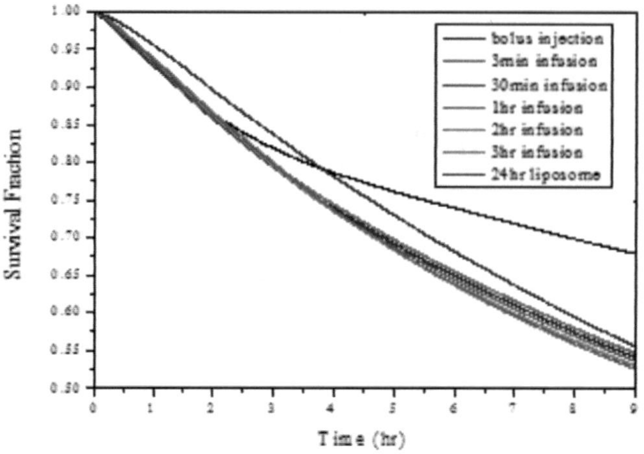

Fig. 2 Predicted anticancer efficacy for different modes of drug administration

(312e) Unraveling the Mechanism of a DNA Nanotechnology: The 10-23 Dnazyme

Margaret C. Linak and Kevin D. Dorfman, Chemical Engineering and Materials Science, University of Minnesota, Minneapolis, MN

Many emerging DNA nanotechnologies are difficult or even impossible to crystallize, frustrating attempts to connect their structure and function. Using a straightforward multi-scale simulation approach, we developed an atomic scale model of one such nanotechnology that has resisted attempts at crystallization, the widely used 10-23 DNAzyme. Deoxyribozymes (DNAzymes) are a new class of small catalytic oligodeoxynucleotides composed entirely of DNA. The 10-23 DNAzyme consists of a conserved 15-base catalytic core that is flanked by binding arms typically 7-10 bases in length with sequence complementary to a target single-stranded RNA molecule. The 10-23 DNAzyme is capable of cleaving at a purine-pyrimidine junction with both high turnover rate and substrate affinity. The ability of the 10-23 DNAzyme to cleave an RNA substrate has lead to both *in vitro* and *in vivo* applications although the evidence for its biochemical activity remains indirect. Our simulations suggest a plausible mechanism for the catalysis of RNA cleavage by this single-stranded DNA molecule. Two critical structural features were found: (i) the unstacking and rotation of the RNA nucleotide at the cleavage site, and (ii) the formation of electrostatic traps for cofactor metal ions.

(312f) Individualized Physiologically Based Modeling and Model Predictive Control of Volatile Anesthesia

Alexandra Krieger, Chemical Engineering, Imperial College London, London, United Kingdom, Nicki Panoskaltsis, Department of Haematology, Northwick Park Hospital, London, United Kingdom, Athanasios Mantalaris, Chemical Engineering and Chemical Technology, Imperial College London, London, United Kingdom, Michael C. Georgiadis, Centre for Process Systems Engineering, Chemical Engineering, Imperial College London, London, United Kingdom and Efstratios N. Pistikopoulos, Chemical Engineering, Imperial College London, Centre for Process Systems Engineering, London, United Kingdom

The modeling and control of anesthesia is believed to benefit the safety of the patient undergoing surgery and provide anesthetists and researchers with valuable insights. It is expected that mathematical modeling and optimized control could (i) pave the way for personalized health care, taking into account the individual patient characteristics for optimal and flexible drug infusion rates, (ii) provide the anesthetist with more time to focus on critical issues, and (iii) guarantee the safety of the patient minimizing side-effects.

This work presents a physiologically based, patient-specific, compartmental model for volatile anesthesia, consisting of a multiple blood and tissue compartmental model, adjusted to the weight, height, sex and age of the patient (Krieger et al. 2011). The predicted hypnotic effect and cardiovascular and respiratory depression is directly correlated to the arterial anesthetic concentration via an effect site concentration and the Hill-equation and (Gentilini et al. 2001). The individual patient characteristics in the pharmacodynamic equations are included via an initial estimate based on the physiologic variables of the patient and updated during the course of anesthesia as a function of the obtained measurements, such as the mean arterial blood pressure.

The derived model is used as the underlying model for model predictive control (MPC) of anesthesia. MPC is the method of choice for the biomedical system of anesthesia, as it can handle multiple-input multiple-output (MIMO) systems and safety constraints on the drug infusion rates, measured physiological variables and drug concentrations. Via applying multi-parametric programming for MPC, a fast and patient specific calculation of the optimal control inputs (Dua & Pistikopoulos 2005), e.g. the drug infusion rates and the inspired concentration of the volatile anesthetic agent is assured.

References

Dua, P. & Pistikopoulos, E.N., 2005. Modelling and control of drug delivery systems. *Computers and Chemical Engineering*, 29, p.2290-2296.

Gentilini, A.L. et al., 2001. Modeling and closed-loop control of hypnosis by means of bispectral index (BIS) with isoflurane. *IEEE Transactions on Biomedical Engineering*, 48(8), p.874-889.

Krieger, A. et al., 2011. A novel physiologically based compartmental model for volatile anaesthesia, *21st European Symposium on Computer Aided Process Engineering*. Elsevier, p. 1495 – 1499

(312g) Computational Model for Nanocarrier Adhesion to Cell Surfaces Validated Using in Vivo, in Vitro, and Atomic Force Microscopy Experiments

Jin Liu, University of Pennsylvania, Philadelphia, PA, Portonovo S. Ayyaswamy, Mechanical Engineering and Applied Mechanics, University of Pennsylvania, Philadelphia, PA, David M. Eckmann, School of Medicine, University of Pennsylvania, Philadelphia, PA, Vladimir Muzykantov, Institute for Environmental Medicine, Department of Pharmacology, University of Pennsylvania School of Medicine, Philadelphia, PA and **Ravi Radhakrishnan**, Department of Bioengineering, University of Pennsylvania, Philadelphia, PA

Our objective is to develop a computational platform for targeting of functionalized nanocarriers to optimize experimental design protocols for drug delivery. A computational methodology based on Monte Carlo and the weighted histogram analysis method has been developed to calculate the absolute binding free energy between functionalized nanocarriers (NC) and endothelial cell (EC) surfaces. The calculated NC binding free energy landscapes yield binding affinities that agree quantitatively when directly compared against analogous measurements of specific antibody-coated NCs (100 nm in diameter) to intracellular adhesion molecule-1 (ICAM-1) expressing EC surface in in vitro cell-culture experiments. The effect of antibody surface coverage (σs) of NC on binding simulations reveals a threshold σs value below which the NC binding affinities reduce drastically and drop lower than that of single anti-ICAM-1 molecule to ICAM-1. The model suggests that the dominant effect of changing σs around the threshold is through a change in multivalent interactions; however, the loss in translational and rotational entropies are also important. We also discuss the role of membrane deformation in mediating the NC adhesion. We validate the simulations against three distinct classes of experiments: invivo targeting in mice, cellular targeting in cultured cells under flow, and binding assays using AFM.

References:

1. A Computational Model for Nanocarrier Binding to Endothelium Validated Using In Vivo, In Vitro, and Atomic Force Microscopy Experiments, J. Liu, G. E. R. Weller, B. Zern, P. S. Ayyaswamy, D. M. Eckmann, V. Muzykantov, R. Radhakrishnan, Proceedings of the National Academy of Sciences, 107(38), 16530-16535, 2010. Pubmed ID: 20823256.

2. Dynamic factors controlling carrier anchoring on vascular cells, T. N. Swaminathan, J. Liu, B. Uma, P. S. Ayyaswamy, R. Radhakrishnan, D. M. Eckmann, IUBMB Life, 2011, 63(8):640-647. Pubmed ID: 21721099

3. Dynamic factors controlling targeting nanocarriers to vascular endothelium, D. M. Eckmann, R. Radhakrishnan, V. Muzykantov, R. Radhakrishnan, D. M. Eckmann, Current Drug Metabolism, 2012, 113, 70. Pubmed ID: 22292809.

4. Multiscale Modeling of Functionalized Nanocarriers in Targeted Drug Delivery, J. Liu, R. P. Bradley, D. M. Eckmann, P. S. Ayyaswamy, R. Radhakrishnan, Current Nanoscience, 2011, 7(5): 727-735. Pubmed ID: 22116782.

5. Multivalent binding of nanocarrier to endothelial cells under shear flow, J. Liu, N. J. Agrawal, A. Calderon, P. S. Ayyaswamy, D. M. Eckmann, R. Radhakrishnan, Biophysical J., 2011, 101(2):319-326. Pubmed ID: 21767483

An Automated Tissue Digester for Pancreatic Islet Production

Zhongliang Lu[a], Paul W. Todd[b], and Thomas R. Hanley[a]
[a]Department of Chemical Engineering, Auburn University, Auburn, AL USA
[b]Techshot, Inc., 7200 Highway 150, Greenville, IN USA

Introduction

Large quantities of pure viable islets are necessary to treat diabetes. Isolation of the islets is the primary step in preparing the islets for the transplantation. Moskalewski[1] in 1965 used collagenase digestion of the pancreas and was the first to obtain large number of islets. Lacy and Kostianovsky[2] in 1967 improved islet yield by cannulation of the pancreatic duct and injection of Hanks solution, which mechanically distended and disrupted the gland before digestion. In 1981, Horaguchi and Merrel[3] described a new technique for the isolation of islets from dog pancreas, but it could not be applied to other larger animals or the human pancreas. Noel et al.[4] in 1982 modified the technique of Horaguchi by injecting collagenase in two separate sections, adding a calcium chelator (EGTA) into the duct. In 1984 Gray et al.[5] modified the previous technique and applied it to the human and monkey pancreas. Warnock et al.[6], from the same group in Edmonton, compared the method of perfusion with single intraductal injection of collagenase. Finally, Ricordi et al.[7] further developed a semi-automated method for the isolation of human pancreatic islets, a definite improvement on existing methods.

In the Ricordi digestion, the pancreas and glass balls are placed into the chamber, creating a closed-loop recirculation and dissociation system operated by intermittent manual shaking of the vessel. However, the yield of islets was still variable (Iwashita, 1996)[8]. One of the most important reasons is the collagenase, essential for islets release from the surrounding tissues, can cause damage to islets by both warm ischemic injury and enzymatic disruption. The problem is further compounded by the variable activation degree of native proteolytic enzymes within acinar cells. Therefore, an automated system that is capable of introducing rigorous standards into the preparation of islets must be established to eliminate the uncertainty. By utilizing an established enzyme digestion system and by creating a reproducible, controllable tissue digester, two of the major variables in islet production are eliminated. The Dynacult reactor has wide application as documentation in micro gravity research (Ivan, 2006)[9]. One advantage is that the fluid flow within the vessel can be carefully controlled, enabling the shear stress experienced by the islets to be optimized when the islets rotate with the fluid flow simultaneously. All above features lay a good foundation for the new digester design. The modification of this reactor was accomplished by introducing wall vanes to enhance turbulence. The details of digester design and subsequent results of FLUENT simulations will be illustrated by diagrams.

Materials and Methods

The design of new digester. The modified Dynacult reactor (MDR) incorporates hemispherical vanes to create turbulence similar to that created by moving glass balls inside the Ricordi chamber. Contrary to Ricordi chamber, the MDR provides a controllable shear rate that allows for a controllable and more favorable digestion environment. In addition, the digestion is automated as the cylinder walls are controlled by motors in contract the manual operation of the Ricordi chamber.

When the bioreactor is operating, the rotation of the core is independent of the exterior wall. The operator may choose to rotate the core at the same rate and in the same direction as the wall, hold it stationary; or rotate it in opposite directions. With the progress of tissue dissociation, the vanes reduce the tissue size and allow maximum islet release and production. As rotational speed increases and the spacing between the walls decreases, the size of the tissue fragments is reduced.

Computational Fluid Dynamics (CFD) Simulation. FLUENT simulates fluid flow by using a discrete-phase Reynolds stress model. The results for both the MDR and the Ricordi chamber are compared. Velocity inlet boundary conditions assume velocities normal to inlet surface, the outlet pressure is set at 0 Pa gauge, and the no-slip condition is assumed at all walls. In addition, FLUENT's preprocessor "Mechanical" is used to generate flow-field meshes. An unstructured mesh consisting of tetrahedral volumetric elements in three dimensions is used in the entire domain with discrete control volumes. The operation mode is optimized by investigating rotating speed, inlet flow rate and vessel dimensions, each separately considered as a variable while others are held constant. The distribution of hydrodynamic numbers and relevant parameters is then displayed by contours generated by CFD-post. Therefore, the optimal operation condition can be determined by comparing all the parameter value and distribution diagrams under different examined conditions.

Experimental Measurements Using Cultisphers. Cultisphers (gelatin bead, size range from 200 to 380 microns) are soaked for two hours in phosphate buffered saline (PBS) prior to use. The particles are rinsed several times and mixed with water to prepare a suspension with known concentration. The motor is operated under four different voltages: 6, 8, 10 and 12 volts. The experiment has two parts. In the input dispersion process, the particle suspension is pumped into the water-filled digester at 60 ml/min while rotating the double walls in opposite directions. The particle motion and collisions are observed. The MDR is operated in a closed circulation at maximum rotation speed. The time interval (dispersion time) is measured between when the particles are first seen and when particles completely fill the vessel. In the output dispersion process, inlet is switched to pure water and the particle elution time is estimated. Particle motion is recorded by eScope digital microscope.

Results and Discussion

MDR configuration and operation modes. The configuration of the MDR is illustrated in Figure 1. The dimensions are listed in Table 1. In order to imitate the turbulent environment caused by glass balls in the Ricordi chamber, the "half-ball baffle" vane was employed to dissociate the pancreatic tissue. In earlier testing all vanes were cemented onto the annular wall evenly distributing four in one circle. The base system, including a stationary base that is mounted to the experiment enclosure and a mobile base that can be repositioned between experiments in order to vary the length of the bioreactor, has been designed to allow the shell, core, and rotational speeds to be changed (Paul, 2004)[10].

Distributions of Hydrodynamic Parameters for the Ricordi chamber. Manual shaking mode is divided into two motions: (1) rolling motion with the rotational axis parallel to central line of the Ricordi chamber at a maximum rotation speed of 66 rpm; (2) simple harmonic motion; velocity formula $V_z = -0.921534 * \sin(16.755161 * t)$ (m/s). Figure 2 shows the distributions of four important hydrodynamic parameters, and Figure 3 displays velocity magnitude contours.

Figure 1 - The Modified Dynacult Digester Design

Table 1 - Dimensions of Tissue Digester

No.	Shell diameter (inch)	Core diameter (inch)	Shell length (inch)	Core length (inch)	Shell shaft (inch)	Core shaft (inch)	Wall distance (inch)	Ball radius (inch)	No. of balls on shell	No. of balls on core	Specific Area (in^{-1})
1	OD=2, ID=1.75	1.5	2.5	1.87	0.362	0.268	0.125	0.083333333	7	5	12.64938126
2	OD=2, ID=1.75	1.25	2	1.875	0	0.125	0.25	0.166666667	3	2	10.79504838
3	OD=2, ID=1.75	1	2	1.3	0.4	0.3	0.375	0.21	2	1	6.827977316
4	OD=2, ID=1.75	1	3	2.37	0.362	0.268	0.375	0.25	3	2	7.460899369
5	OD=2, ID=1.75	1	3	2.969	0	0.031	0.375	0.27	3	2	8.536585366
6	OD=2, ID=1.75	1	3.5	2.85	0.372	0.278	0.375	0.29	3	2	7.865438369
7	OD=2, ID=1.75	1	3.953	3.375	0.237	0.341	0.375	0.3	3	2	7.826278815
8	OD=2, ID=1.75	0.5	2	1.3	0.4	0.3	0.625	0.125	4	2	4.536291251

- **Distributions of Cell Reynolds Number:**

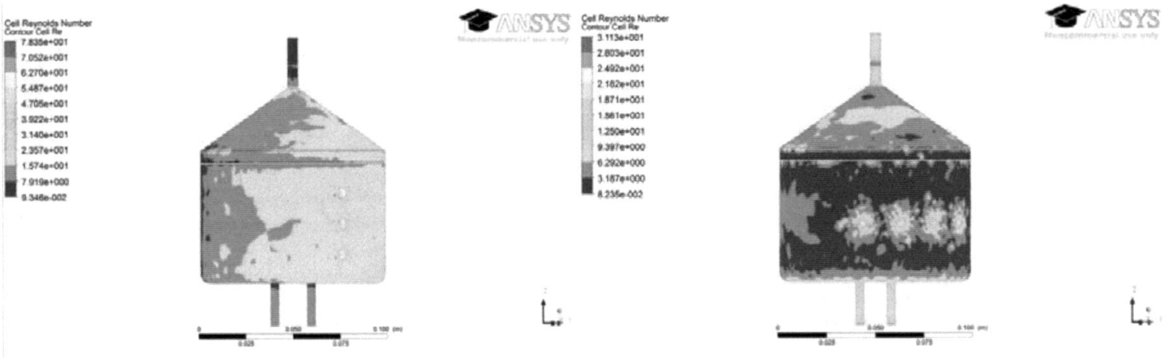

- **Distributions of Shear Rate (s^{-1}):**

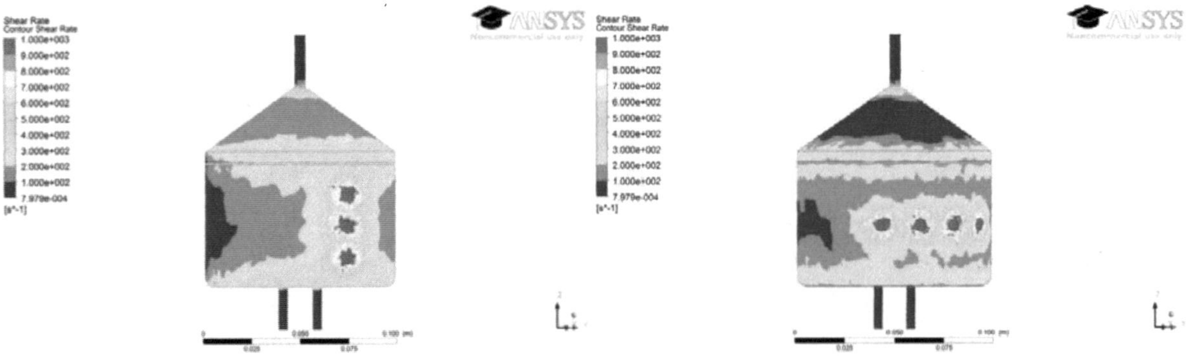

■ **Distributions of Energy Dissipation Rate per Volume (W/m^3):**

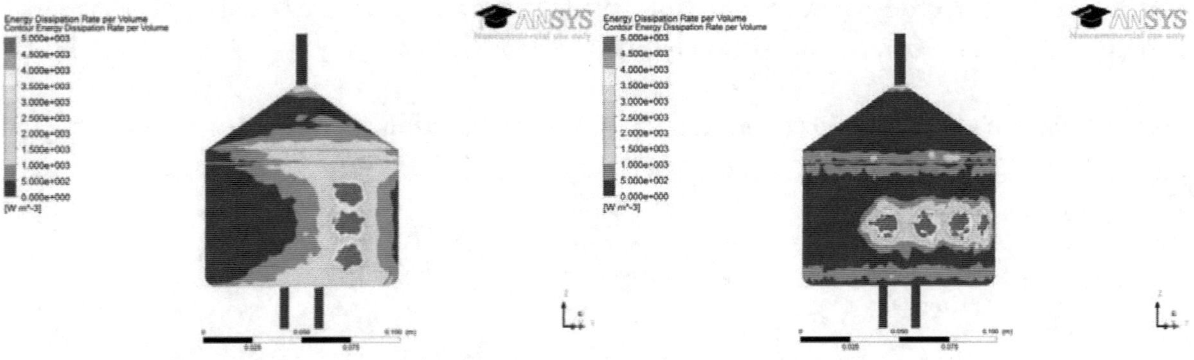

■ **Distributions of Kolmogoroff Length(m):**

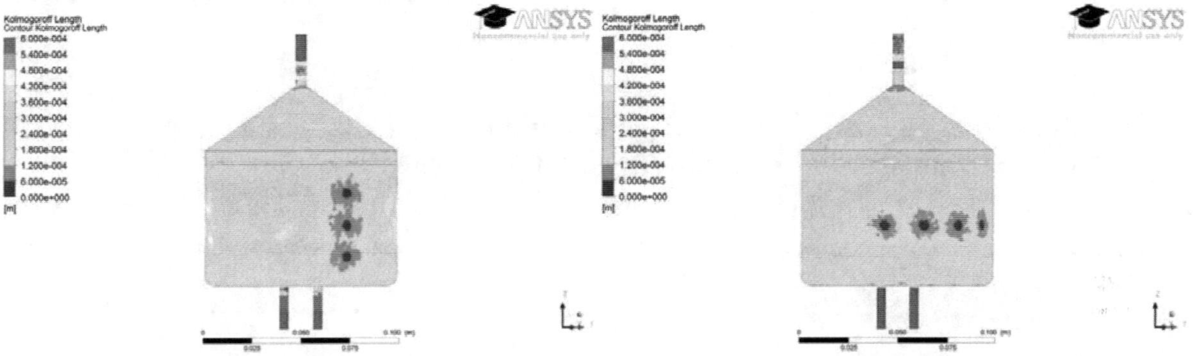

Figure 2 - The contour of hydrodynamic parameters for the chamber.
(left column - rolling motion; right column - up and down motion)

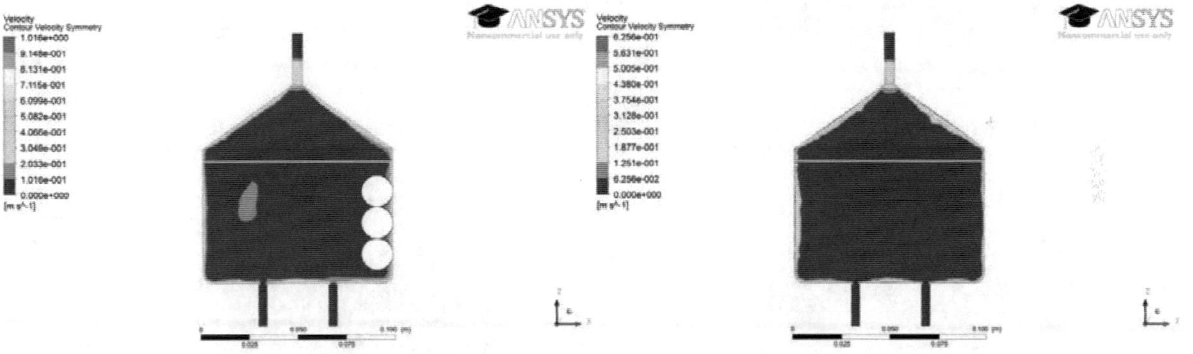

Figure 3 - The velocity magnitude contours on symmetric plane.
(left column - rolling motion; right column - up and down motion)

Distributions of hydrodynamic parameters for MDR. The rotation speed, inlet flow rate and vessel dimensions were investigated independently; the double wall rotation position is assumed to have impact on shear environment. Based on FLUENT simulation results, the preferred 100 rpm rotation speed is firstly determined through comparing the shear rate and energy dissipation rate per volume at the same inlet flow rate and dimension. It is also found that the cylinder wall rotation has little impact on flow parameters so that all the contours can be displayed at a single position with minimum vane spacing. Subsequently, the optimal inlet flow rate 120 ml/min and vessel dimension 3 (see Table 1) are obtained by similar comparison under corresponding variable gradients. Totally,

35 combinations of triple variables are verified by FLUENT simulation and displayed by CFD-post. The distributions of hydrodynamic parameters, velocity magnitude contours and particle pathline under optimal operation condition are shown in Figure 4 and 5, respectively.

- **Distributions of Cell Reynolds Number:** **Distributions of Shear Rate (s^{-1}):**

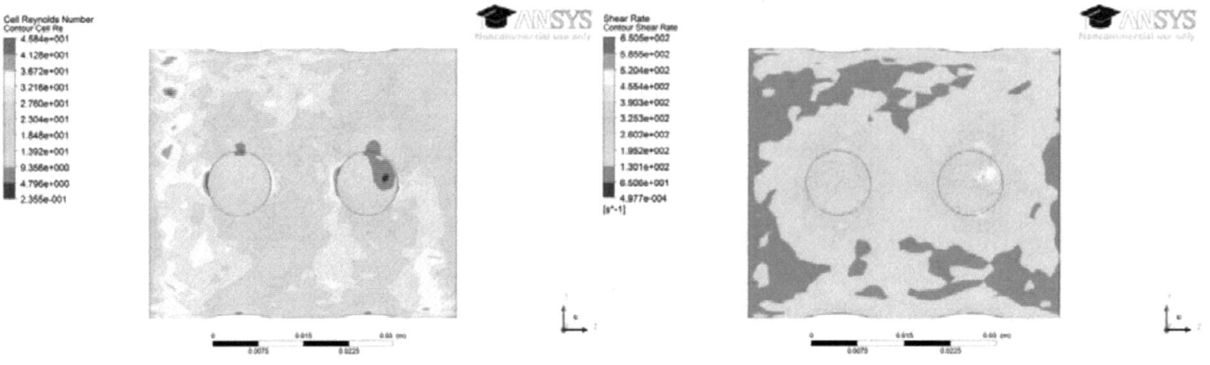

- **Distributions of Energy Dissipation Rate per Volume(W/m^3) (Left column below)**
- **Distributions of Kolmogoroff Length(m) (Right column below)**

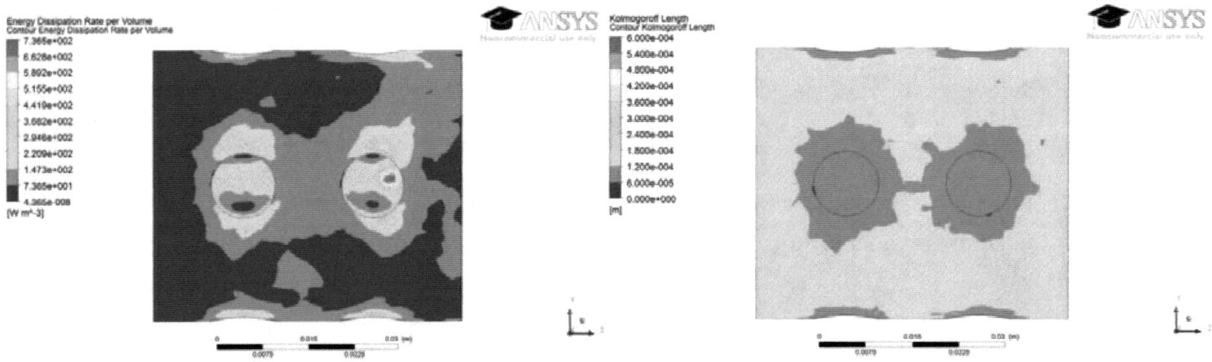

Figure 4 - The contour of hydrodynamic parameters for the MDR

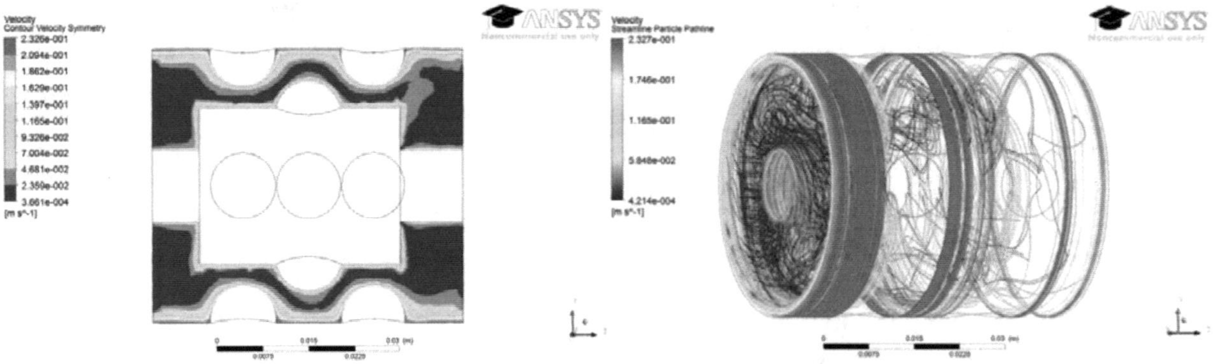

Figure 5 - The velocity magnitude contours on symmetric plane (Left)
The particle pathline (unit: m/s) shown in the MDR (Right)

Comparison of results between Ricordi chamber and MDR. Three flow parameters, including shear rate, shear stress, and energy dissipation rate per volume, are summarized as volume average, minimum and maximum value in Table 2, respectively.

Table 2 - Numerical simulation results for Ricordi chamber and MDR

Flow parameters \ Digester type		Ricordi chamber		MDR	
		Rolling motion	Up & down motion	A position	B position
Shear Rate (s^{-1})	VolAve	28.28	17.61	49.5	48.08
	Min	0.000797941	0.00219062	0.000497681	5.781E-05
	Max	4360.54	3363.51	650.549	620.068
Shear Stress (pa)	VolAve	0.4874	0.131	2.279E-01	2.141E-01
	Min	2.402E-05	1.420E-05	3.959E-06	3.969E-06
	Max	14.3727	12.956	2.23194	1.26812
P/V (W/m^3)	VolAve	46.3	20.52	18.9	17.41
	Min	1.92E-08	3.61E-07	4.36E-08	2.061E-08
	Max	57879.7	30878.6	736.461	754.62

P/V - energy dissipation rate per volume; VolAve - volume average value of corresponding parameters. Spacing between two adjacent vanes is minimum at A , maximum at B

The comparison of above parameters indicate that volume average shear stress and P/V for the MDR are lower than that for Ricordi chamber, with lower maximum values as well. Moreover, the maximum of shear rate has the same results as that of shear stress and P/V. However, volume average of shear rate shows the opposite results.

The General Description of Particle Motion and the Mixing Process in the MDR. An experimental analysis of the relative particle-to-fluid motion under the gradient of rotation speed shows that the particle will move along a certain path controlled by the action of different hydrodynamic forces. The particle dispersion/ residence time and relevant phenomenon are summarized in Table 3. The particle distribution is snapshotted using the eScope digital microscope under 12.5 magnification, both in stationary and rotating state (Figure 6). The effect of opposite rotation of the cylinder walls on the particles tracks is illustrated. Increasing the rotating speed of the inner wall will cause particles to migrate toward the outer wall. The greater the rotating speeds of

Table 3 - The Dispersion Process for Cultisphers

Voltage (V)	RPM	dispersion mixing time (s)	Input phenomenon	Residence time (min)	Output phenomenon
6, shell only	7	22	Most particles are sedimentary.	null	null
8, shell only	10	19	Sedimentary and suspended particles are nearly equal.	null	null
10, shell only	13	15	More particles are suspended.	null	null
12, shell only	17	13	More particles are suspended with less and scattered sedimentary ones.	null	null
12, in opposite directions	17&21 (shell & core)	11	No sedimentary particles are left	10~11	Particles are in radial symmetry distribution; turbulent flow

the outer wall, the less eccentric rate of the curve and the greater migrating rate of the particle in radial direction. Also, the final trajectory of the particle is closer to the inner cylinder. Fully developed turbulent flow can be seen through the transparent shell wall. The complete mixing process occurs in the intermediate region between adjacent vanes and double walls without any sedimentary particles while the particles near the walls will move with the shell/core in the same direction. The particles collide with each other in two directions due to radial and axial motion.

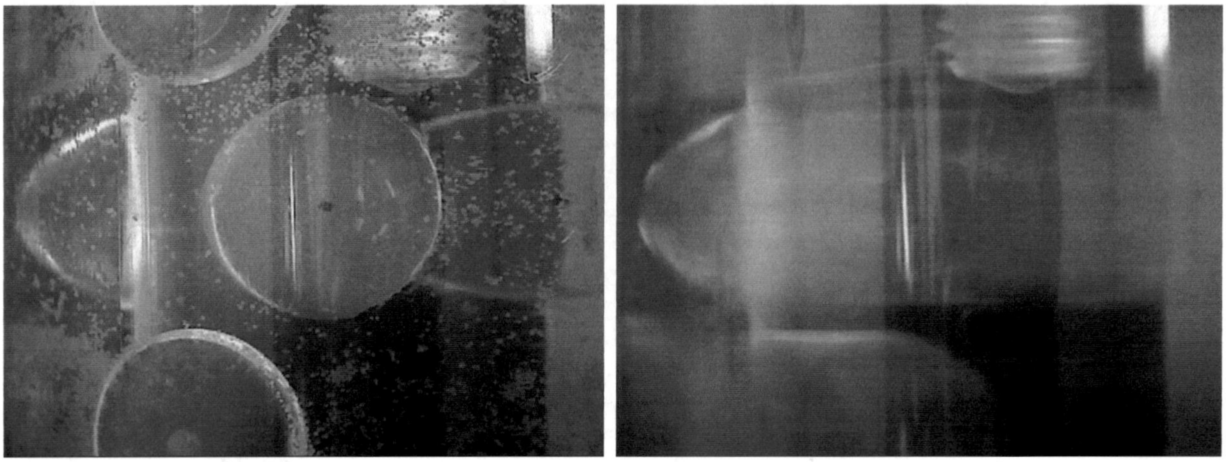

Figure 6 - Image from eScope microscope using Cultisphers as tissue fragment surrogates. Left: particles suspending in stationary state; Right: particles moving with walls rotating

The Potential Problems and Alternative Strategy for MDR Design. Existing technologies of pancreatic digestion have fundamental limitations. The design of the MDR, including the spacing between vanes and the rotational rates of the outer and inner modules, may need to be studied in greater details if we are unable to either suspend tissue during the digestion or if we are unable to separate islets from acinar tissue. The complex interaction between the tissue fragments and the MDR walls/vanes may require additional analysis resulting in a redesign. Problems with this design may require additional CFD analysis in order to estimate the shear stress and the convection forces in the MDR. The preliminary design of interacting hemispherical vanes may be altered. We also may only allow one of these modules to implement in order to simplify the simulation.

Conclusions

The new design of the MDR digester succeeding partial feature of Ricordi chamber overcomes the deficiency of manual operation. This digester configuration is designed to enhance the turbulent effect and maximize contact area with more probability of collision and mixing between tissue fragments and vanes. FLUENT simulations confirm a more uniform and lower shear environment in the MDR that should produce more viable pancreatic islets. The values of the shear and energy dissipation rates at the ball contact area in the Ricordi chamber are significantly larger than values in the MDR, resulting in an expected loss islet viability. The volume average value of shear rate for the Ricordi chamber is slightly lower than that for the MDR. Ricordi chamber fluid away from walls is near stationary. Comparing MDR results between rotation position A and B indicates the cylindrical wall motion has little effect on the shear and energy dissipation rates.

References

1. Moskalewski, S. (1965), “Isolation and Culture of the Islets of Langerhans of the Guinea Pig,” *Gen. & Comp. Endocrinology*, **5**: 342-353.
2. Lacy, P. E., and M. Kostianovsky (1967), “Method for the Isolation of Intact Islets of Langerhans from the Rat Pancreas,” *Diabetes*, **16**: 35-39.
3. Horaguchi, A. and R. C. Merrell (1981), “Preparation of Viable Islet Cells from Dogs by a New Method,” *Diabetes*, **30**: 455-458.
4. Noel, J., A. Rabinovitch, A, L. Olson et al. (1982), “A Method for Large Scale High-Yield Isolation of Canine Pancreatic Islets of Langerhans,’ *Metabolism*, **31**: 184-187.
5. Gray, D. W. R., P. McShane, A. Grant and P. J. Morris (1984), “A Method for Isolation of Islets of Langerhans from the Human Pancreas, *Diabetes*, **33**, 1055-1061.
6. Warnock, G. L., D. Ellis, R. V. Rajotte et al. (1988), « Studies of the Isolation and Viability of Human Islets of Langerhans, *Transplantation*, **45**, 957-963.
7. Ricordi C, P. E. Lacy, E. H. Finke et al. (1988), « Automated Method for Isolation of Human Pancreatic Islets, *Diabetes*, **37**, 413-420.
8. Iwashita, C., T. Asano, T. Kenmochi, K. Jingu, T. Uematsu, T. Nakagohri, M. Hasegawa, M. Maruyama, H. Miyauchi, and K. Isono (1996), “Combined Method of Mechanical Chopper and Automated Two-Step Digestion Technique for Islet Isolation from Canine Pancreas, *Transplantation Proceedings*, **28(1)**, 337.
9. Rivera-Solorio, I., and, S. J. Kleis (2006), “Model of the Mass Transport to the Surface of Animal Cells Cultured in a Rotating Bioreactor Operated in Micro gravity,. *Biotech. & Bioengr.*, **94(3)**, 496.
10. Todd, P. W., L. Hodapp, A. Jones, D. Kennedy, T. Popson and , E. Taylor (2004), “Robotic Bioreactor for Cell Culture in Space Applications (Dynacult). Final Progress Report,”. 1-119.

(312i) An Integrated Computational Model of Powder Release, Dispersion, and Deposition in a Dry Powder Inhaler

Jovana Milenkovic, CERTH/CPERI, Thessaloniki, Greece, Aleck Alexopoulos, Chemical Process Engineering Research Institute, Thessaloniki, Greece and Costas Kiparissides, Department of Chemical Engineering, Aristotle University of Thessaloniki & Chemical Process Engineering Research Institute, Thessaloniki, Greece

Dry Powder Inhalers, DPIs, are the principle means of delivering pharmaceuticals due to their ease of use and cost-effectiveness. The main function of a DPI device is the adequate dispersion and delivery of particles. Initially the particles are in the forms of a loose powder which, under the action of airflow, is broken up and dispersed as particle aggregates which are then further broken up into fine particles. Powder properties, e.g., cohesion, charge, size, and size distribution, influence powder dispersion and the breakage of particle agglomerates. One of the problems with DPIs is the loss of powders due to deposition within the device which ranges from 20-40% of the total drug. In order to provide the maximum drug dose per inhalation and to ensure proper function of the DPI device for each dose (i.e., dose-to-dose variation) it is desirable to minimize the drug loses due to internal deposition. It is also desired to have good control over the dispersibility of the powder, release of drug (if attached to powder particles), and breakup of agglomerates in order to achieve the desired specification of particle/agglomerate size distributions at the exit flow of the DPI. The Turbuhaler (AstraZeneca) is a multidose dry powder inhaler that is widely used to deliver a number of drugs (typically for asthma) to the upper respiratory tract. Each dose is initially in the form of loosely packed drug aggregates, ~10-20μm in size, which are released into a mixing/dispersion chamber, where they are broken up into particles, ~1-2μm in size, which are then directed to the inhalation channel of the device (Tsima et al., 1994; Wetterlin, 1988).

In this work the airflow in a Turbuhaler DPI is determined by computational fluid dynamics, CFD, simulations. The DPI geometry is first constructed in a CAD/CAM environment (i.e., CATIA V5 R19) and then imported into GAMBIT (v2.1) where a tetrahedral grid is constructed. The computational grids consisted of 1-2 10^6 tetrahedral cells with a maximum skewness of 0.85. The equations for flow were solved using commercial CFD software (i.e., FLUENT v6.3). The SST k-ω model for transient turbulent flows was employed. Different steady state airflows were simulated by imposing a wide range of pressure drops at the mouthpiece outflow of -200 to -10000Pa corresponding to airflow rates of 2.7 and 18.6 l/min, respectively. Dynamic airflow simulations were performed by imposing a transient pressure boundary condition at the mouthpiece channel. Eulerian/Lagrnagian simulations of particle motion and deposition were conducted for particles between 1-20μm in size corresponding to the single particle and particle agglomerate size ranges of the powder. Particles and particle aggregates are assumed to be released instantaneously and uniformly from a surface located upstream from the powder storage site. For dynamic simulations the particles are released either instantaneously or at a constant rate over a period of time < the inhalation time.

The computed airflow in the DPI revealed a strongly non-uniform flow field. Particle depositions occurred mostly in the dispersion chamber and the helical region of the DPI where strong tangential motion was observed. The computed pressure drops were in agreement to available experimental data. In order to compute realistic particle depositions in the device a nonideal sticking efficiency, determined from particle/wall adhesion models, was employed. The predicted total particle depositions were in agreement to available literature data. This work represents the first coupling of particle adhesion models with fluid dynamics simulations for inhalation devices and permits the investigation of the effect of particle formulation properties on the DPI outflow and internal loses.

M.P. Tsima, G.P. Martin, C. Marriott, D. Gardenton and M. Yianneskis, 1994. Drug delivery to the respiratory tract using dry powder inhaler, International Journal of Pharmaceutics, 101 () 1-13.

K. Wetterlin, 1988 Turbuhaler:A New Powder Inhaler for Administration of Drugs to the Airways. Pharmaceutical Research, Vol, 5, No.8.

(408a) Stimulus-Responsive Controlled Release System by Covalent Immobilization of Proteins Into Mesoporous Silica Nanoparticles

Kai H. Griebenow, Chemistry, University of Puerto Rico, Rio Piedras, San Juan, PR

Modern treatment options in cancer and other therapies increasingly rely on passive and active targeting of the drug. In this context, mesoporous silica nanoparticles (MSN) have emerged as an attractive class of drug delivery carriers for therapeutic agents. We recently explored the covalent immobilization of model proteins into MSN to generate stimulus-responsive controlled release systems. The model proteins carbonic anhydrase (CA) and cytochrome c (Cyt-c) were coupled to sulfosuccinimidyl 6-[3'(2-pyridyldithio)-propionamido]hexanoate (Sulfo-LC-SPDP) and subsequently covalently immobilized into MSN via thiol-disulfide interchange. The model proteins could be released from MSN with 10 mM glutathione which represents intra-cellular redox conditions while it remained bound to the MSN at extra-cellular redox conditions represented by 1 μM glutathione. Bioactivity of the model proteins was maintained well during immobilization and release. Studies using human cervical cancer (HeLa) cells showed good biocompatibility. MSN containing immobilized Cyt-c induced apoptosis in HeLa cells. We will also highlight recent data on endocytosis of Cyt-c loaded MSN and endosomal escape obtained by confocal microscopy. Furthermore, we will present data on active targeting using MSN by binding of of suitable ligands.

(408b) Engineering and Characterizing Aggregation-Resistant Antibodies

Joseph M. Perchiacca, Shantanu V. Sule, Ali Reza A. Ladiwala, Jayapriya Jayaraman, Moumita Bhattacharya and Peter M. Tessier, Dept of Chemical & Biological Engineering, Rensselaer Polytechnic Institute, Troy, NY

Protein aggregation is one of the most significant challenges in successfully developing and delivering monoclonal antibodies for therapeutic applications. We are investigating how to engineer the sequences of antibodies to promote high solubility without altering biological activity, as well as how to identify optimal formulation conditions that maximize antibody solubility using high-throughput measurements of antibody self-association. In this presentation, we will first discuss our work on investigating the impact of the complementarity determining regions (CDRs) of antibodies on their aggregation behavior for a library of homologous antibody fragments (Perchiacca et al., Proteins, 2011). Importantly, we find that charged mutations near the edges of hydrophobic CDR loops significantly increase antibody solubility without reducing binding activity. Second, we will discuss our development of a high-throughput method (self-interaction nanoparticle spectroscopy, SINS) for measuring antibody self-association (Sule et al., Biophys J, 2011). We find that SINS can reliably and rapidly identify sequence variants and formulation conditions that minimize monoclonal antibody aggregation.

(408c) Characterization of Monoclonal Antibody Conformations and Self-Associations At High Concentrations Using Neutron Scattering Techniques

Yun Liu[1,2], Eric J. Yearley[1], Isidro Zarraga[3], Norman J. Wagner[1], Thomas M. Scherer[3], Steven J. Shire[4] and Yatin R. Gokarn[3], (1)Chemical and Biomolecular Engineering, University of Delaware, Newark, DE, (2)Center for Neutron Science, NIST, Gaithersburg, MD, (3)Late Stage Pharmaceutical and Processing Development, Genentech Inc., South San Francisco, CA, (4)Pharmaceutical R&D, Genentech Inc., South San Francisco, CA

Monoclonal antibodies (MAb) have become a crucial therapeutic agent. However, some applications such as a subcutaneous injection need highly concentrated protein solutions that can have undesired large viscosity. Therefore, it is of great importance to examine the relation between MAb self-associated nanostructure and its viscosity under various concentration and excipient conditions with techniques which can probe both structure and dynamics that span the length scale of an individual protein to a larger aggregate. We have investigated the protein-protein interactions and diffusive properties of highly concentrated MAbs using small angle neutron scattering (SANS) and neutron spin echo (NSE). Our results indicate that concentration, temperature, pH and surfactant do not have a strong effect on the individual MAb conformation for our MAb samples. The short-range attraction of some MAb proteins is found to be highly anisotropic in contrast to many other protein solutions at low ionic strength. Using neutron spin echo, we have accurately measured the short-time self-diffusion coefficients of MAbs at very large concentrations. One of MAb proteins is found to form small clusters at a wide range of concentrations. We speculate that the cluster formation is thus the underlying cause of the dramatic increase in viscosity as a function of concentration in this specific MAb protein.

(408d) Aggregation of Proteins Studied by Deep UV Resonance and Nonresonance Raman Spectroscopy

Sergey Arzhantsev[1], Connie Ruzicka[1], Vincent Vilker[2] and John Kauffman[1], (1)Division of Pharmaceutical Analysis, US FDA, Saint Louis, MO, (2)Office of Testing and Research, FDA, Silver Spring, MD

Aggregation of proteins is one of the degradation processes which occurs during the storage of pharmaceutical protein products, and can significantly decrease the potency of protein drugs. Several studies have shown that aggregation correlates with changes in the secondary structure of proteins. Raman spectroscopy is a nondestructive analytical method that is sensitive to the secondary structure of proteins. Deep UV resonance Raman (DUVRR) spectroscopy has the potential to be a good alternative technique to UV and vibrational circular dichroism for determination of secondary structure in formulated therapeutic proteins. The resonance excitation increases the Raman signal, specifically enhancing the intensity of peptide vibration bands, and completely removing fluorescence background. The advantage of DUVRR spectroscopy is that it does not usually require any sample preparation and is as sensitive as circular dichroism. We have developed a mathematical algorithm for mixture decomposition in order to remove excipient spectral contributions from the total Raman signal of formulated protein products. This presentation will discuss the use of DUVRR spectroscopy to study the aggregation of insulin and insulin variants, and will compare the capabilities of resonance Raman and nonresonance Raman spectroscopies.

(408e) Reactivity of Polymersome Encapsulated Hemoglobin with Physiologically Important Gaseous Ligands: Oxygen, Carbon Monoxide, and Nitric Oxide

Shahid Rameez[1], **Uddyalok Banerjee**[2], Jorge Fontes[3], Alexander Roth[2] and Andre Palmer[2], (1)Dept. of Biochemistry, The Ohio State University, Columbus, OH, (2)Department of Chemical and Biomolecular Engineering, The Ohio State University, Columbus, OH, (3)Dept. of Chemical and Biomolecular Engineering, The Ohio State University, Columbus, OH

Two distinct preparations of amphiphilic diblock copolymer vesicles (i.e., polymersomes), composed of (poly-(ethylene oxide)−poly(butadiene)) (PEO−PBD), with molecular weights of 1.8 and 10.4 kDa, offering different hydrophobic membrane thicknesses, were used to encapsulate the oxygen (O2) storage and transport protein hemoglobin (Hb) for possible application as a red blood cell (RBC) substitute. Key biophysical properties like p50 and cooperativity as well as the kinetics of polymersome encapsulated Hb (PEH) interaction with physiologically important gaseous ligands (O2, carbon monoxide, and nitric oxide) were measured as a function of the hydrophobic membrane thickness of the PEH particle. Taken together, the results of this work show that PEHs exhibit biophysical properties and retarded ligand binding/release kinetics (compared to cell-free Hb), which are similar to the behavior of RBCs. Therefore, PEHs have the potential to serve as safe and efficacious RBC substitutes for use in transfusion medicine. Detailed studies involving different amphilic diblock copolymers can be conducted in future to explore these possibilities.

(408f) Low-Viscosity Highly Concentrated Dispersions of Stable Protein Nanoclusters for Subcutaneous Injection

Aileen K. Dinin[1], Ameya U. Borwankar[2], Maria Andrea Miller[1], Tarik A. Khan[3], Brian Wilson[1], Kevin Kaczorowski[4], Jennifer A. Maynard[5], Thomas M. Truskett[6] and Keith P. Johnston[7], (1)Dept. Chemical Engineering, University of Texas at Austin, Austin, TX, (2)Chemical Engineering, University of Texas at Austin, Austin, TX, (3)Department of Chemical Engineering, The University of Texas at Austin, Austin, TX, (4)Chemical Engineering, Massachusetts Institute of Technology, Cambridge, MA, (5)Chemical Engineering and Materials Science, University of Minnesota, Minneapolis, MN, (6)Chemical Engineering and Institute for Theoretical Chemistry, The University of Texas at Austin, Austin, TX, (7)Chemical Engineering, The University of Texas, Austin, TX

Stabilizing proteins at high concentration is of broad interest in drug delivery, for treatment of cancer and many other diseases. Proteins have a tendency to undergo irreversible aggregation, gelation or precipitation at high concentrations due to unfolding caused by specific short ranged forces. Herein, we create highly concentrated antibody dispersions (up to 260 mg/ml) comprising of dense equilibrium nanoclusters of protein [monoclonal antibody (mAb) 1B7, polyclonal sheep Immunoglobulin G (IgG) and bovine serum albumin (BSA)] molecules, which upon dilution *in vitro* or administration *in vivo,* remain conformationally stable and biologically active. The nanoclusters are formed by adding trehalose as a cosolute which strengthens the short-ranged attraction between protein molecules. The protein cluster diameter was reversibly tuned from 50 to 300 nm by balancing short-ranged attraction against long–ranged electrostatic repulsion of weakly charged protein at a pH near the isoelectric point (pI). The high protein volume fraction within the cluster stabilizes the protein conformation through a self-crowding mechanism and the primarily repulsive electrostatic interactions between the clusters keep them colloidally stable, preventing gelation. Upon dilution of the dispersion *in vitro*, the clusters rapidly dissociated into fully active protein monomers as shown with biophysical analysis (SEC, dynamic light scattering (DLS), circular dichroism (CD) and SDS-PAGE) and sensitive biological assays. *In vivo* subcutaneous injection into mice results in indistinguishable pharmacokinetics *versus* a standard antibody solution. This tunablility of multi-scale interactions , specifically inter-monomer short range attraction and inter-cluster long range repulsion, to produce nanoclusters is not protein-specific. Therefore this nanocluster concept can be applied to a wide range of therapeutic proteins without the need to engineer modified proteins through changing the amino acid sequence. Stable protein dispersions with low viscosities may potentially enable patient self-administration by subcutaneous injection of antibody therapeutics being discovered and developed.

(451a) Building Tools for Predicting Allosteric Regulation Pathways in Proteins

Galen Collier and Vanessa Ortiz, Chemical Engineering, Columbia University, New York, NY

Allosteric regulation of protein function is a mechanism by which a structural event occurring at one site within a protein's overall structure causes an effect at another site, and allostery plays a central role in a huge variety of mechanistic pathways in biological systems. The prediction and modulation of allosteric responses in proteins holds promise for applications in a diverse range of research areas, including drug design and materials engineering. The structural changes involved in protein allostery include deformations where local elastic moduli play a major role. Knowledge of the elastic modulus at each atom position or within specific regions in a protein can provide valuable information about the local mobility and stability under conformational deformations. However, most methods used for calculating local elastic moduli cannot be applied to proteins in a straightforward manner because they require second derivatives of the potentials that describe the interactions amongst a protein's atoms. To address this issue, we have begun the development of computational methods that enable the measurement of elastic moduli at different positions in a protein in order to build a 3D map of local mechanical properties that can be used to identify pathways of allostery in proteins and protein networks. Presented here are the results from a preliminary set of protein structural analyses completed using these methods to calculate elastic moduli in an atomistic fashion.

(451b) Computer Aided Design of Bio-Reactor for Bone Tissue Engineering

Marina Campolo, Center for Fluid Dynamics and Hydraulics, University of Udine, Udine, Italy, Francesco Curcio, Dept. Biological and Medical Sciences, University of Udine, Udine, Italy and Alfredo Soldati, Department of Mathematics and Informatics, University of Udine, Udine, Italy

In this work we propose a modelling approach to identify sets of culture conditions to promote homogeneous growth of cells in perfusion bioreactors equipped with regular shape scaffolds.
Homogeneous growth is the outcome of an initial homogeneous cell seeding and of an homogeneous cell proliferation which depend on a number of variables including pH, temperature, oxygen tension, nutrient supply, catabolites removal and also biochemical and mechanical stimuli. Consequently, the entire reactor and its operating protocols should be carefully designed to achieve a homogeneous distribution of cells during seeding, to maintain conditions which promote homogeneous proliferation and to prevent conditions which may lead to cell necrosis.
Providing the right mechanical stimulation and adequate nutrient supply are key issues in cell culturing: at optimal operating conditions the level of shear stress should be large enough to stimulate cell proliferation/differentiation (Porter et al., 2005, Ma et al., 2007, Olivares et al., 2009) and, yet, not too large to cause cell damage or detachment from the scaffold; in addition, nutrient feeds should be large enough to replenish continuously species consumed by cellular metabolic activity and removal of catabolites should be efficient enough to avoid local accumulation of species (e.g. gas bubbles) which may affect the culture microenvironment.
We use a zero dimensional mass transport and reaction model to identify cases in which dynamic culturing is necessary to avoid nutrient limitation. Then, based on the three dimensional rendering of the flow field inside the bioreactor, we (1) identify regions where cellular growth may become critical due to the low/high value of local shear stress and (2) we evaluate the risk of accumulating catabolites in the form of gas bubbles which may lead to clogging of scaffold microchannels; finally, using a one dimensional mass transport and reaction model we calculate the minimal perfusion flow necessary to maintain the cellular growth rate above a target threshold and we identify the conditions (bio-reactor region, flow rate, limiting bubble size) for which bubbles may adhere stably to the scaffold, potentially leading to bubble clogging.
The developed approach is used to analyse culturing conditions inside an indirect perfusion bioreactor equipped with a lattice scaffold. Regions where the perfusion flow is inadequate to foster cellular growth at the desired rate are identified and the probability of local clogging is evaluated. The perfusion flow required to maintain the target growth rate inside the bioreactor is calculated.

(451c) Modelling and Robust Model Predictive Control of Insulin Delivery in Type 1 DM

Stamatina Zavitsanou[1], Nicki Panoskaltsis[2], Athanasios Mantalaris[1], Michael C. Georgiadis[1] and Efstratios N. Pistikopoulos[3], (1)Department of Chemical Engineering, Centre for Process Systems Engineering, Imperial College, London, United Kingdom, (2)Department of Haematology, Northwick Park Hospital, London, United Kingdom, (3)Chemical Engineering, Imperial College London, Centre for Process Systems Engineering, London, United Kingdom

The determination of effective strategies to treat Type 1 Diabetes Mellitus is of paramount importance since it can be characterised as one of the most prevalent severe chronic diseases in childhood. Motivated by the challenge to ensure the patients' safety and to improve their quality of life the development of a closed loop system of optimal insulin delivery has been the most widely studied approach in the field (Kovatchev et al., 2010), (Grosman et al., 2010). In this framework, a detailed physiologically based compartmental model is presented, extended from previous studies (Zavitsanou, et al. 2011). It represents personalised glucose-insulin interactions in the fasting, prandial and postprandial state while insulin is administered exogenously with an insulin pump and mimics the mechanism of pancreatic response. Glucose coming from food consumption and endogenously produced is distributed in blood and tissue compartments, representing the involved organs and glucose regulation is achieved with the appropriate subcutaneous insulin infusion to compensate for high blood glucose concentrations. Patient specific characteristics such as age, gender, weight, height, hematocrit and insulin sensitivity are included and the model can represent a virtual patient. The individual insulin delivery (dosing and scheduling) is optimised in terms of maintaining blood glucose concentration within the safe range. Closed loop studies in the context of explicit/multi-parametric model predictive control (Dua et al., 2006) are presented to evaluate the performance of the controller that takes into account safety constraints of desired blood glucose concentration and operational constraints of available insulin delivery. Case studies of closed loop insulin delivery of patients subjected to external disturbances are presented to highlight the challenges of the system such as high intra patient variability and time delays.

Dua, P., Doyle, F. J., & Pistikopoulos, E. N. (2006). Model-based blood glucose control for Type 1 diabetes via parametric programming. *IEEE transactions on bio-medical engineering*, *53*(8), 1478-91.

Grosman, B., Dassau, E., Zisser, H. C., Jovanovič, L., & Doyle, F. J. (2010). Zone model predictive control: a strategy to minimize hyper- and hypoglycemic events. *Journal of diabetes science and technology*, *4*(4), 961-75.

Kovatchev, B., D, P., Cobelli, C., Renard, E., Anderson, S., Breton, M., et al. (2010). Multinational study of subcutaneous model-predictive closed-loop control in type 1 diabetes mellitus: summary of the results. *Journal of diabetes science and technology*, *4*(6), 1374-81.

Zavitsanou, S., Panoskaltsi, N., Mantalaris, A., Georgiadis, C, M., & Pistikopoulos, E. N. (2011). Modelling of the Insulin Delivery System for patients with Type 1 Diabetes Mellitus. *21st European Symposium on Computer Aides Process Engineering*.

(451d) Deriving 2D Velocity Profile Using Streamlines Image Velocimetry (SIV)

Eliezer Keinan[1], Elishai Ezra[1] and Yaakov Nahmias[1,2], (1)Bioengineering Program, Benin School of Computer Science and Engineering , The Hebrew University of Jerusalem, Jerusalem, Israel, (2)Department of Cell and Developmental Biology, Silberman Institute of Life Sciences, The Hebrew University of Jerusalem, Jerusalem, Israel

Introduction

Fluid dynamics plays a fundamental role in environmental, biological, and industrial processes ranging from mass transport to separations and heat exchange. Investigation of sand dunes, wind waves, heat transfer, fouling, and aneurysm failure are a few examples where an exact solution of the fluid velocity field is required. Current methods to investigate this velocity field are cumbersome and hence, a rapid, accurate and user friendly analysis tool is required.

Current direct velocimetry methods include Particle Imaging Velocimetry (PIV), Particle Tracking Velocimetry (PTV), and Laser Doppler Velocimetry (LDV). These methods require significant investment in equipment and computational power. For example, a commercial PIV system consists of a laser diode and an air-cooled superfast camera.

Here, we introduce a novel method termed Streamline Image Velocimetry (SIV) which allows a rapid quantification of velocity fields in arbitrary geometries. We use a single long-exposure image to capture multiple particles streaking in individual streamlines across the field. Streamlines confine streamtubes, in which the volumetric flow is constant for incompressible fluid. Using the explicit analytical solution as an initial condition, we quantify the velocity field throughout the entire field. We compiled our results into a computer program that enables the rapid derivation of velocity profiles from a single input image, which can be taken using a simple household camera. Our results show an excellent correlation to numerical simulations and are able to quantify fluid velocities across complex porous barrier, where numerical simulations are difficult at best.

Results

Analytical Solution for Rectangular Channel

The analytical solution for the velocity field in the middle of a low aspect ratio (h<<w) rectangular channel is approximated as a Fourier series in equation 1:

$$(1)\ V_x(y)=48Q/(\Pi^3*(w-0.63h))*\Sigma^{\infty}_{n,odd}*n^{-3}*(1-\cosh(n\Pi y/h)/\cosh(n\Pi w/2h))*\sin(n\Pi/2)$$

Where v_x is velocity vector, h is the channel height, w is the channel width, y is horizontal location inside the channel and Q is the volumetric flow rate.

Streamline Imaging

To experimentally quantify the fluid velocity field we introduced fluorescent microbeads into a microfabricated channel which contained distortions. Long exposure images were taken using a fluorescent microscope, taking care to trace complete paths of multiple particles across the entire field of view. The average velocity between two streamlines is defined by volumetric flow rate divided by stream tube cross section. Polynomials were fitted to the visualized streamlines. A mean polynomial was calculated between every two neighboring streamlines. The velocity direction was determined by the derivative of the mean polynomial. The distance between the streamlines was calculated as the length of a line perpendicular to the mean polynomial drawn from the upper polynomial to the lower one.

Computational Image Analysis

The image processing and velocity profile calculation were compiled into a single MATLAB program (code is available on request) with a graphical user interface (GUI). The program enables to load streamlines snapshot, preliminary image processing (alignment, cropping, binary conversion, PSF deconvolution). The program is loaded with macroscopic parameters, as fluid viscosity, density and "anchor" region geometry. Based on the entered parameters, the program can fit a set of polynomials to the streamlines, and calculate the velocity field. Data can be presented as either numerical values, or an arrow field.

SIV Analysis of Flow in Microfluidic Device

We fabricated a PDMS device with a negative arched distortion and took snapshots of the streamlines using fluorescent microscopy. We analyzed the streamlines snapshots in MATLAB and calculated the velocity profile in cross sections. The velocities were compared to COMSOL derived numerical simulations: in the "anchor" region where the program calculates the analytical solution, in the middle of the arch where the streamlines are parallel and there is no perpendicular velocity component and in the quarter of the arch. The results show high correlation (0.97). The correlation between SIV and the numerical calculation is high (0.96) even at the quarter of the distortion where the perpendicular component of the velocity is maximal presenting a rapid change in velocity as function of distance from the channel wall.

SIV Analysis of a Microfluidic Channel with Porous Elements

In order to verify our method with a different flow irregularity, characterized with unknown boundary conditions, we produced the velocity field of a visualized streamlines snapshot, taken from a recent publication. The visualized streamlines were taken from microfluidic device integrated with nanoporous forests of vertically aligned carbon nanotubes. In such devices the complexity of the velocity field calculation increases due to the porous boundary conditions. However, for SIV, the boundary conditions are the streamtubes enabling the derivation of the velocity field based solely on the visualized streamlines. The results indicate that SIV produces high-precision results in complex flow regions where the fluid velocity magnitude and direction rapidly changes. SIV analysis presenting velocimetry at low cost and reduces computational resources.

Materials and Methods

Chemicals

Poly(dimethyl)siloxane (PDMS) was purchased from Dow Corning (SYLGARD). SU8-2050 was purchased from Microchem (Newton, MA). Fluorescent green microbeads, Pluronics F68, and Phosphate Buffered Saline (PBS) were purchased from Sigma Aldrich (Rehovot, Israel).

Numerical Modeling

Numerical simulations were performed using COMSOL multiphysics simulation platform v3.5A with a direct linear system solver and extra fine mesh (PARADISO). Fluid density was defined as $1x10^3$ kg/m^3, with a dynamic viscosity of $1x10^{-3}$Pa·s.

Microfluidic Channels

Microfluidic devices were fabricated by soft lithography. Briefly, molds were fabricated by photolithography of SU8 on silicon wafers at the Hervey Krueger Center of Nanoscience and Nanotechnology at the Hebrew University of Jerusalem. Channels were replica molded in PDMS and bonded to glass using oxygen plasma bonding. The device is 100 μm high. Main channel width is 3000 μm and the disturbance depth is 100 μm.

Streamline Imaging

Prior to the experiment the device channel was coated with Pluronic F-68 for 1 hour at room temperature to prevent non-specific adhesion of fluorescent particles. The experiments were performed with microbeads diluted in 1% Pluronic F-68 at a concentration of $2 \cdot 10^6$ particles/ml. Microbeads were perfused using a Chemyx Fusion 200 syringe pump and imaged on a Zeiss Axiovert LSM700 Microscope.

Streamlines Image Analysis

Images processing was carried out in MATLAB. Image were imported and deconvolved using Gaussian filter generated from the theoretical point spread function (PSF). Streamlines were traced using MATLAB 2011b algorithm. Polynomial functions were fitted to traced streamlines and transformed into a binary 2D matrix.

Streamlines Image Velocimetry

A streamline is a curve parallel to the velocity vector which can be visualized using long exposure snapshots of fluorescent microparticles in fluid. In steady flow, the streamlines coincide with the paths of the fluid particles. A streamtube is a surface confined between streamlines. The volumetric flow rate in a streamtube is conserved; therefore, the product of the streamtube cross-section (A) and the average velocity (w) is constant as described in Equation 2:

$$(2)\ A_1 w_1 = A_2 w_2 = Aw$$

Our method relies on the fact, that in a streamtube that is formed in a simple geometry-shaped vessel, the volumetric flow rate is conserved when entering into a more complex geometry and hence, the volumetric flow rate in the complex region can be derived from flow in the simple geometry. Hence, our method requires that the analyzed vessel will include a simple geometry fluid entrance region, which will serve as an "anchor" for further calculations. Our analytical solution is suitable for rectangular cross section channels, however it can be easily modified for any other geometry.

(451e) Non-Dimensional Analysis of Retinal Microaneurysms: Critical Threshold for Treatment

Elishai Ezra[1], Eliezer Keinan[1], Yossi Mandel[1], Michael Boulton[2] and Yaakov Nahmias[1,3], (1)Bioengineering Program, Benin School of Computer Science and Engineering, The Hebrew University of Jerusalem, Jerusalem, Israel, (2)Department of Anatomy and Cell Biology, University of Florida, (3)Department of Cell and Developmental Biology, Silberman Institute of Life Sciences, The Hebrew University of Jerusalem

Introduction: Diabetic retinopathy is responsible for 5% of the cases of blindness worldwide and for up to 17% of the cases in the western world. Retinal microanuerysms reduce vision quality due to local loss of endothelial barrier function, causing leakage and retinal edema. Localized leakage from microaneurysms can be detected and blocked using laser ablation, slowing the progression of diabetic blindness. Our approach has been to gain a critical understanding of the fluid dynamics of the problem, which will enable us to predict risk associated with the diseased morphology. However, analytical solutions of Navier-Stokes equations governing fluid dynamics exist solely for restricted conditions, while numerical simulations can only be preformed on one set of initial and boundary conditions at a time. We derived a non-dimensional expansion parameter from an approximate analytical solution and generalized it using an experimentally validated numerical solution, uncovering a non-dimensional behavior, which allows for rapid calculation of critical properties. Such properties could be used to rapidly calculate changes in fluid velocity, shear force, or fluid pressure due to microvessel expansion, without the need for complex analytical or numerical solutions. In order to validate our findings, we used high-resolution confocal images of diabetic retinas and observed that microaneurysms above our critical threshold demonstrated a dramatic, size-dependent increase in von Willebrand Factor (vWF) expression, a marker of endothelial dysfunction.

Materials and Methods: Numerical simulations were performed using COMSOL multiphysics simulation platform v3.5A with a direct (PARADISO) linear system solver and extra fine mesh. The experimental validation of the numerical model was performed in microfluidic devices, which were fabricated by soft lithography at the Harvey Krueger Center of Nanoscience and Nanotechnology at the Hebrew University of Jerusalem. Postmortem donor eyes from diabetic patients with early retinopathy characterized by the presence of microaneurysms were obtained from the National Disease Research Interchange (Philadelphia, USA). Retinas were examined from the vitreous surface using Biorad MRC 600 confocal laser scanning attachment mounted on a Zeiss Axiovert microscope using a 25 mW argon laser (Ion Laser Technology). Maximum projection images were analyzed using the Image J software (NIH) to quantifying microaneurysm geometry and average vWF intensity.

Results and Discussion: Endothelial dysfunction occurs when shear force drops below 0.4 Pa, enabling us to calculate a critical non-dimensional expansion ε of 0.38. Our analysis sets a critical aneurysm diameter of 15 μm for capillaries, 25 μm for venules and 104 μm for arterioles, above which endothelial dysfunction is predicted to occur. To test this hypothesis we obtained high-resolution confocal images of human diabetic retinas and quantified microaneurysm's size distribution. The percentage of microaneurysms above our threshold ranged from 10% to 60%. Remarkably, while vWF expression in microaneurysms below our threshold was not significantly different than the surrounding microvasculature (p=0.076), it was 60±10% lower in microaneurysms above our threshold ($p=7\times10^{-15}$).

In contrast to the high correlation between vWF to the non-dimensional expansion, there was a weak correlation with the aneurysm size (R2=0.59). The rapid advance of optics over the past decade permits high-resolution imaging of the retina in vivo with functional markers, such as vWF. A combined imaging of structure and function can lead to optimization and customization of preventive treatment before the onset of retinal edema. Our method will allow experts to identify and treat microaneurysms posing a high-risk of leakage, prior to edema, minimizing damage and saving vision.

Conclusions: Our approach has been to gain a critical understanding of the fluid dynamics in retinal microaneurysms, which will enable us to predict the risk associated with the diseased morphology by uncovering a non-dimensional behavior, allowing rapid calculation of critical propermmmDa sdsdly LE in the quarter i while tified ties.

(451f) Engineered Macromolecules As Inhibitors to Oxidized Low Density Lipoprotein by Macrophage Scavenger Receptors: Simulation of Structure – Function Relationships

Michael Tomasini, Biomedical Engineering, Rutgers University, Piscataway, NJ and M. Silvina Tomassone, Chemical and Biochemical Engineering, Rutgers University, Piscataway, NJ

Atherosclerosis begins when high levels of circulating low-density lipoproteins (LDL) are transported into the extravascular space and become trapped in the extracellular matrix where they are subject to oxidative modification. Oxidized LDL causes damage to surrounding endothelial cells and initiates an inflammatory signaling cascade which results in the recruitment of monocytes to the site of injury. Monocytes differentiate into macrophages that express surface scavenger receptors active in the non-specific uptake of modified LDL. The excessive uptake of oxidized LDL results in the formation of foam cells which become engorged, die, and subsequently form atherosclerotic plaques. One possible method to combat plaque formation is through the use of inhibitors to prevent uptake of oxidized LDL by the scavenger receptors. Tian et al. 2004 [1] developed a class of micelle-forming molecules termed nanolipoblockers (NLB) comprised of a mucic acid head group with attached aliphatic chains and a long polyethylene glycol (PEG) tail used to block uptake of oxidized LDL by scavenger receptors. In this work, we use coarse-grained molecular dynamics simulations to explore the structural and electrostatic properties of NLBs with the scavenger receptor A (SR-A) to determine optimal interaction and thus inhibition of oxidized LDL uptake.

We use a coarse-grained model based on the MARTINI force field [2] to describe the NLB molecules as well as the collagen-like domain of SR-A containing the binding pocket for oxidized LDL [3]. We include four different NLB structures varying the charge, charge location, and the NLB ability to form micelles. We find that for NLB molecules containing aliphatic groups near the mucic acid head, micelles form within 300 ns with a micelle diameter of 7.6 – 12.8 nm. As an indirect assessment of binding affinity, we identified the number of contacts that each type of NLB made with SR-A both at 0.8 nm and 2.0 nm from the receptor. The number of close contacts (< 0.8 nm) were similar for each NLB, but aligned with experimental trends showing anionic micelle-forming NLBs as the best inhibitors of oxidized LDL uptake. In calculating the specific contacts between the charged portions of NLBs and the binding pocket of SR-A indicated that the addition of a negative charge on the PEG tails of NLBs did not act to enhance interaction with the important residues of the SR-A, however an anionic charge at the NLB head increased interaction with the positively charged residues of the binding pocket, confirming experimental findings

References:

[1] Tian L, Yam L, Zhou N, Tat H, and Uhrich KE. Amphiphilic Scorpion-like Macromolecules: Design, Synthesis, and Characterization. *Macromolecules*. 2004; **37**: 538-543.

[2] Monticelli L, Kandasamy SK, Periole X, Larson RG, Tieleman DP, and Marrink SJ. The MARTINI Coarse-Grained Force Field: Extension to Proteins. *J. Chem. Theory and Comput.* 2008; **4**(5): 819-834.

[3] Kodama T, Freeman M, Rohrer L, Zabrecky J, Matsudaira P, Krieger M. Type I macrophage scavenger receptor contains alpha-helical and collagen-like coiled coils. *Nature*. 1990; **343**(6258): p. 531-535.

(451g) Mechanistic Modeling of DNA Hybridization On Surfaces for Improved Microarray Design

Kyle E. Pratt, Ryan C. Welling, Terry J. Schmitt and Thomas A. Knotts IV, Department of Chemical Engineering, Brigham Young University, Provo, UT

DNA microarrays are chip-based, analysis tools which can perform hundreds of thousands of assays in parallel. They can be used to determine the identity of genes or gene expression levels present in a sample and have been identified as a key technology in genomic sciences and emergent medical techniques. Despite their abundant use in research laboratories, microarrays have not been used in the clinical setting to their fullest potential due to the difficulty of obtaining reproducible results. Microarrays work on the principle of DNA hybridization, and can only be as accurate as this process is robust, yet fundamental, molecular-level understanding of hybridization on surfaces is lacking. Specifically, a detailed knowledge of the various factors affecting hybridization of probes to targets is needed to further refine these devices.

Using a carefully-parameterized, coarse-grain model of DNA, we have spent the last several years studying the thermodynamics of hybridization on surfaces for simple probe/target pairings. This has led to several important insights, but a detailed mechanism of the process has never been reported. In this presentation, we outline efforts to fill in several gaps in understanding. A detailed description of the hybridization process is presented first. These results were obtained using extensive computer simulations to create a projection of the free energy of hybridization onto two reaction coordinates: the probe/target separation distance and the angle of approach between the two strands. Several cases, such as strands of equal and unequal lengths in the bulk and on the surface, are presented. The free-energy projections were obtained from two-dimensional umbrella sampling, and each required over 11,000 simulations. Though computationally demanding, this work has created an unprecedented picture of the hybridization process.

In addition to the mechanisms described above, we also present results on how factors found in real microarrays affect the hybridization process. Specifically, we show how the presence of multiple probes on the surface affects hybridization, and describe how the number and location of mismatched base pairs affect the stability of the duplex. Finally, we describe how the hybridization process is influenced by the locations of probe/surface connections. As a whole, the results give the most complete picture to date of the biophysics involved in microarray performance and how this knowledge can be used to improve next-generation devices.

(451h) Modeling Nonspecific Interactions in Biological Systems

Andrew D. White, Ann K. Nowinski, Wenjun Huang, Andrew J. Keefe, Fang Sun and Shaoyi Jiang, Chemical Engineering, University of Washington, Seattle, WA

Many of the difficulties in biomedical applications arise from the complexities of multiple interactions in biological systems. These interactions can be broadly broken into two categories: those which are important to function (high activity for a specific target) and nonspecific binding which is detrimental to function (low activity binding to many targets). Nonspecific interactions are central to many failures of drugs, devices and materials in the biomedical field. Despite this fact, there is little modeling work on studying nonspecific interactions. Modeling all possible interactions within a biological system is difficult, yet there are ways to both indirectly study nonspecific binding and directly study many possible interactions using bioinformatics. In this work, we describe recent work using bioinformatics, phenomenological modeling, molecular simulations, and experiments to study three types of nonspecific interactions. They are the weak nonspecific interactions between proteins in crowded environments, irreversible nonspecific protein adsorption, and nonspecific stabilization of proteins in molecular chaperones. These disparate systems have allowed us to generalize our conclusions into a small set of principles that other researchers should consider when designing specific activity in order to maintain resistance to nonspecific interactions.

(451i) Exploring the Relationship Between Helicobacter Pylori's Caga Sequence and Affinity with Host's Receptors: A Proposal for Molecular Diagnostic Tools

Paula Andrea Delgado Pinzón[1], Carlos Alberto Jaramillo Henao[2], María del Pilar Delgado Perafán[2], Natalia Melisa Peñaranda Fajardo[2], Nathalia Garces Ferrreira[3], Harold Enrique Castro Barrera[3] and Andrés Fernando González Barrios[1], (1)Chemical Engineering, Universidad de los Andes, Bogotá, Colombia, (2)Biological Sciences , Universidad de los Andes, Bogotá, Colombia, (3)Systems and Computing Engineering , Universidad de los Andes, Bogotá, Colombia

Helicobacter pylori is a human pathogen reported to be present in nearly 50% of world's population[1,2]. Moreover the virulence of the strain is one of the major determinants for disease developing among infected people. One of the major factors associated to *H. pylori* virulence is the cytotoxin associated gene A (*cag*A) which encodes the CagA protein[3,4]. This protein is injected directly from the microorganism through type 4 secretion system (T4SS) to gastrointestinal epithelial cells, where disrupts the cellular signaling mechanism[5]. It is known that CagA has a 5' region highly conserve while 3' region is variable due to the presence of different types and/or numbers of repeat sequences which contains an EPIYA motif location where the phosphorylation of the protein takes place[2,6]. These differences in sequence may have a relationship with the affinity of CagA with host's receptors and hence with the pathogenicity of *H. pylori*. Here in this work we obtained several sequences of different Colombian CagA strains and determine the tertiary structure using I-Tasser[7] server based on multiple threading complemented with Replica Exchange Dynamics technique[8] using opportunistic Cloud Infraestructure. Then, these structures were docked[9,10] with different receptors including SCHP-2 and Src. Free energy calculations were correlated with the pathogenicity of the strains which were previously categorized by its pathogenicity level (cancer, chronic ulcer, chronic gastritis etc) in order to determine a potential tool for molecular diagnostics.

References

1. Higashi H. SHP-2 Tyrosine Phosphatase as an Intracellular Target of *Helicobacter pylori* CagA Protein. Science. 2001;295(5555):683-686.

2. Acosta N. *Helicobacter pylori* CagA protein polymorphisms and their lack of association with pathogenesis. World Journal of Gastroenterology. 2010;16(31):3936.

3. Backert S, Tegtmeyer N, Selbach M. The versatility of *Helicobacter pylori* CagA effector protein functions: the master key hypothesis. Helicobacter. 2010;15(3):163-176.

4. Jones KR, Whitmire JM, Merrell DS. A Tale of Two Toxins: *Helicobacter pylori* CagA and VacA Modulate Host Pathways that Impact Disease. Frontiers in Microbiology. 2010;1.

5. Handa O, Naito Y, Yoshikawa T. CagA protein of *Helicobacter pylori*: A hijacker of gastric epithelial cell signaling. Biochemical Pharmacology. 2007;73(11):1697-1702.

6. Salehi Z, Jelodar MH, Rassa M, Ahaki M, Mollasalehi H, Mashayekhi F. *Helicobacter pylori* cagA Status and Peptic Ulcer Disease in Iran. Digestive Diseases and Sciences. 2008;54(3):608-613.

7. Roy A, Kucukural A, Zhang Y. I-TASSER: a unified platform for automated protein structure and function prediction. Nature protocols. 2010;5(4):725-738.

8. Sugita Y, Okamoto Y. Replica-exchange molecular dynamics method for protein folding. Chemical Physics Letters. 1999;314(1-2):141-151.

9. Qu K, Brooijmans N. Structure-Based Drug Design Computational Methods for Protein Structure Prediction and Modeling. In: Xu Y, Xu D, Liang J, eds: Springer New York; 2007:135-176.

10. Pierce B, Weng Z. Structure prediction of protein complexes. Computational methods for protein structure prediction and modeling. 2007:109-134.

(453a) Improving Stem Cell Transplantation Through Fluid Dynamics and Polymer Physics

Sarah C. Heilshorn, Materials Science and Engineering, Stanford University, Stanford, CA

Stem cell transplantation is a promising therapy for a myriad of debilitating diseases and injuries; however, current delivery protocols are inadequate. Transplantation by direct injection, which is clinically preferred for its minimal invasiveness, commonly results in less than 5% cell viability, greatly inhibiting clinical outcomes. We demonstrate that a significant cause of cell death occurs during transport through the syringe needle, where extensional stretching forces experienced by the cells results in mechanical membrane disruption. We further demonstrate that a variety of clinically relevant cell types including adipose-derived stem cells (ASC), marrow stromal cells (MSC), and neural progenitor cells (NPC) can be protected from these damaging forces by encapsulation within hydrogels of specific viscoelastic properties. We hypothesize that these shear-thinning hydrogels may undergo plug-flow through the syringe needle, whereby bands of shear-thinned polymer are located at the needle walls and lubricate the flow of individual gel "plugs" through the needle. These gel plugs may serve as transport vehicles for the encapsulated cells, thereby shielding them from disruptive mechanical forces.

Building on these fundamental studies, we have designed a reproducible, bio-resorbable, customizable hydrogel using protein-engineering technology. In our Mixing-Induced Two-Component Hydrogel (MITCH), network assembly is driven by specific and stoichiometric peptide-peptide binding interactions. By integrating protein science methodologies with polymer physics models, we manipulate the polypeptide chain interactions and demonstrate the direct ability to tune the network crosslinking density, sol-gel phase behavior, and gel mechanics. These MITCH materials enable stem cell and growth factor encapsulation upon simple mixing at constant physiological conditions, making them well suited for use in the surgical suite. *In vivo* studies in a murine model demonstrate that transplanted ASCs achieve significantly better retention at the subcutaneous injection site compared to cells delivered in collagen, alginate, or saline alone. These results provide mechanistic insight into the role of mechanical forces during cell delivery and support the use of protective hydrogels in future clinical stem cell injection studies.

(453b) 2D Culture and 3D Material Arrays to Define Optimal Conditions for Tenogenic MSC Differentiation

Steven R. Caliari[1], Emily A. Gonnerman[1] and Brendan A. Harley[1,2], (1)Chemical & Biomolecular Engineering, University of Illinois at Urbana-Champaign, Urbana, IL, (2)Institute for Genomic Biology, University of Illinois at Urbana-Champaign, Urbana, IL

The repair of tendon injuries is critically limited by the inability to produce large numbers of tendon fibroblasts, or tenocytes. Multipotent cells, such as mesenchymal stem cells (MSCs), are a potential cell source for tendon repair and can be driven to differentiate down osteogenic, chondrogenic, adipogenic, and other lineages using established protocols. While previous work has highlighted the possible roles of mechanical stimulation, cell confinement/alignment, and co-culture with primary tenocytes, there is no standard protocol for robust tenogenic MSC differentiation. One reason for this is the lack of molecular markers to distinguish tendon from other mesenchymal tissues. Recently, scleraxis (SCX) and thrombospondin-4 (THBS4) have been suggested as tendon-specific markers. Here we present 2D and 3D culture systems to evaluate combinations of biochemical and biophysical factors to drive tenogenic MSC differentiation.

We first used a 2D culture system to assess the effects of a wide range of biochemical factors on the expression of tenogenic markers SCX and THBS4. Human MSCs were seeded in 96-well plates at a density of 1×10^3 cells/well and cultured for 14 days in the presence of single or binary combinations of tenogenic soluble factors bFGF, IGF-1, GDF-5, and GDF-7. Additionally, the influence of standard differentiation agents dexamethasone and ascorbic acid was tested. MSCs were fixed, incubated overnight with primary antibodies against SCX and THBS4, tagged with fluorescently labeled secondary antibodies, and then assayed on a spectrophotometer. Fluorescence was normalized to cell number (DAPI). We found that the presence of dexamethasone and ascorbic acid resulted in increased levels of both SCX and THBS4 on a per cell basis for nearly every soluble factor combination assayed. Additionally, the IGF-1 and GDF-7 groups elicited the greatest increase in SCX and THBS4 production per cell.

In parallel, we have developed arrays of 3D collagen-GAG (CG) scaffolds. Tenocytes are known to de-differentiate in standard 2D culture so we used this system to interrogate the combined effect of scaffold microstructure and biochemistry on tenogenic MSC differentiation. Scaffold arrays with two distinct microstructural regions were fabricated by controlling local heat transfer during the freeze-drying process. A specialized mold was created for this purpose, consisting of a polysulfone chip (2 mm thick) with circular holes (6.5 mm diameter) mounted on a removable base. The geometry/spacing of the nodes was designed to be identical to the dimensions of a 96-well plate. The removable base contained an aluminum section and a polysulfone section ($k_{aluminum}/k_{polysulfone} \sim 900$); this disparity in thermal conductivity was intended to control local heat transfer, and subsequently ice crystal growth kinetics, during freezing. The mean pore sizes for scaffolds were 88 ± 28 μm within the aluminum section and 157 ± 47 μm within the polysulfone section of the array for a freezing temperature of -40°C. Cells seeded in the scaffold array and stained with calcein showed no significant differences in fluorescence compared with cells seeded in a standard 96-well plate. Within the array we could also pattern near-linear ($R^2 = 0.91$) step-wise gradients of a model protein (biotinylated concanavalin A) using carbodiimide chemistry. Ongoing work with these systems is continuing to define optimal biochemical and biophysical cues for guiding tenogenic MSC differentiation.

(453c) Effect of Early Endoderm Induction On Late Stage Pancreatic Maturation of Differentiating Human Embryonic Stem Cells

Maria Jaramillo[1], Shibin Mathew[2] and Ipsita Banerjee[2], (1)Depatment of Bioengineering, University of Pittsburgh, Pittsburgh, PA, (2)Chemical and Petroleum Engineering, University of Pittsburgh, Pittsburgh, PA

Pancreatic differentiation of mouse and human embryonic stem cells has received considerable attention over the last decade. While there has been some success in deriving insulin positive cells from embryonic stem cells, typically the differentiated cells are limited in yield and functionality. Most of the successful differentiation protocols involve a step-wise directed differentiation strategy starting with definitive endoderm (DE) induction; followed by pancreatic progenitor induction and finally islet maturation. While extensive studies have established multiple pathways of endoderm induction, the effect of these pathways on late stage pancreatic maturation remains elusive. The aim of this study is to develop a thorough understanding of the effect of pathway of early endoderm induction on late stage pancreatic maturation.

The human embryonic stem cells were differentiated to mature insulin producing cells following a directed differentiation approach. The first stage of definitive endoderm was induced following four alternative pathways, based on most promising reports by different groups. Namely, we used Activin A in combination with one of the following molecules and growth factors: Fgf2, Bmp4, Wnt3a or PI3K Inhibitor. DE derivatives obtained from these four pathways were subjected to same maturation protocol: pancreatic progenitor induction using sonic hedgehog inhibition and retinoid signaling; islet maturation through notch inhibition. The differentiation potential of each of the endoderm derivatives were compared at different stages of differentiation.

Our results clearly indicate that late stage pancreatic maturation is distinctly governed by the initial pathway of endoderm commitment. From the results obtained from 6 independent studies, we found that DE induction by treatment with PI3KI resulted in at least two times higher upregulation of DE markers than any other conditions; however, at later stages of differentiation these cells obtained a moderate 800-fold upregulation of insulin. Conversely, cells induced towards DE by treatment with Wnt3a and Fgf2 showed between 10,000 and 20,000 fold upregulation of insulin. Consistent with PCR analysis, Wnt3a led to highest yield of C-peptide expressing cells with around 25% positive cells. Interestingly, while the Bmp4 derived DE cells resulted in only about 2-fold increase in insulin expression, they showed the highest upregulation of glucagon: around 4,000 times higher than undifferentiated cells, leading to the conclusion that inducing DE with Bmp4 leads to more alpha cells than beta cells. Detailed analysis using computational methods showed Wnt3a and Fgf2 cultures to show high similarity in gene patterning, and Wnt3a to follow closer resemblance to pancreatic organogenesis in respect to pancreatic marker progression.

We conclude that using Activin A in combination with Wnt3a at the definitive endoderm stage leads to the highest insulin upregulation as well as higher yield of C-peptide expressing cells. Moreover, our analysis shows that mature cell type specification is affected by early differentiation events. Namely our results suggest that while Wnt3a derivatives differentiate into beta-like cells, the use of Bmp4 preferentially directs cells toward alpha cell commitment. To the best of our knowledge, no other study compares methods of definitive endoderm induction and their effect at later stages of development. This study is important since it provides a better understanding of pathways involved in early events of pancreatic differentiation and allows for standardization of these stages in order to achieve better yield and functionality of differentiated human embryonic stem cells

(453d) Directing Pluripotent Stem Cell Differentiation Using Hydrogel Microspheres

Samuel S. Chang, Auburn University, Auburn, AL, Alexander J. Hodge, Chemical Engineering, Auburn University, Auburn, AL and **Elizabeth A. Lipke**, Department of Chemical Engineering, Auburn University, Auburn, AL

Pluripotent stem cells are a promising cell source for treatment of diseased or damaged tissues, including repairing the human heart. Current pluripotent stem cell differentiation protocols generally focus on deriving cardiac tissue with high purity and function, yet few of these methods are highly scalable and also tightly control the cellular microenvironment. Using a water-in-oil emulsion technique, this research has created a scalable and easily modifiable hydrogel microsphere system for pluripotent stem cell differentiation. Microspheres supported the proliferation of the encapsulated pluripotent stem cells and subsequent cardiac differentiation. MESCs were successfully encapsulated into the PEG-fibrinogen microspheres and maintained for over 23 days. mESCs proliferated and initiated formation of embryoid bodies (EBs) within the microspheres during the first week of differentiation. It was found that proliferation and EB formation occurred more readily when cells were encapsulated at high densities (45 and 60 million cells/mL precursor solution) rather than low densities (below 15 million cells/mL precursor). Cell encapsulation concentration did not have a significant impact on initial mean sphere diameter. EBs typically degraded the microsphere and attached to the culture vessel within 10 days of differentiation; these attached tissues initiated spontaneous contraction as early as Day 11 of differentiation. qPCR analysis of gene expression confirmed the down-regulation of pluripotent marker Oct4 and the up-regulation of Nkx2.5. Frequency of spontaneous contraction ranged from 1.5 to 2 Hz.

(453e) Analysis of the Proliferation Potential of Mesenchymal Stem Cells As a Function of Potency with a High-Capacity Clonal Assay

Kim OConnor[1], Katie Russell[1], Michelle Lacey[2], Alan Tucker[3] and Donald Phinney[4], (1)Chemical and Biomolecular Engineering Dept., Tulane University, New Orleans, LA, (2)Department of Mathematics, Tulane University, New Orleans, LA, (3)School of Medicine, Tulane University, New Orleans, LA, (4)Department of Molecular Therapeutics, The Scripps Research Institute, Jupiter, FL

Human mesenchymal stem cells (MSCs) from bone marrow are a heterogeneous ensemble of progenitors and lineage-committed cells, with a broad range of regenerative properties. Ex vivo expansion to produce sufficient quantities of MSCs is essential for most therapeutic applications. The present study resolves the relationship between proliferation potential of MSCs and their potency. Clonal analysis with a unique high-capacity assay generated single-cell derived colonies of MSCs that were classified according to their trilineage potential to exhibit adipo- (A), chondro- (C), and osteogenesis (O) as a measure of potency. Multipotent OAC clones were highly proliferative with colony-forming efficiencies that ranged from 35% to 90%; whereas, O clones formed colonies with an efficiency of 5% or less ($P<0.01$). Similar trends were evident during ex vivo expansion: for example, the median specific growth rate was 0.85 per day (20 h doubling time) for cultures inoculated with OAC clones and was 5-fold less for inocula of O clones ($P<0.01$). OA and OC clones had similar proliferation potentials. More than 75% of cells in subconfluent cultures inoculated with O clones stained positive for senescence-associated beta-galactosidase activity vs. less than 10% for OAC clones ($P<0.001$). Apoptotic cells were in the minority for all potency groups. Preliminary data generated during clonal analysis suggest that osteogenic potential of MSCs to produce mineralized matrix is a function of potency, as well. These results are discussed in the context of the preparation of efficacious MSC therapies by ex vivo expansion.
This research was sponsored by grants to Prof. O'Connor from the National Institutes of Health (R03EB007281) and Tulane University.

(453f) Hydrogel Systems to Examine Diffusion-Mediated Paracrine Signaling On Hematopoietic Stem Cell Fate

Bhushan Mahadik, Sara Pedron, Paul J.A. Kenis and Brendan A.C. Harley, Chemical and Biomolecular Engineering, University of Illinois at Urbana-Champaign, Urbana, IL

Hydrogel systems to examine diffusion-mediated paracrine signaling on hematopoietic stem cell fate

Bhushan Mahadik[*], Dr. Sara Pedron[*], Dr. Paul Kenis[*,#], Brendan Harley[*,#]

[*]Dept. of Chemical and Biomolecular Engineering, University of Illinois at Urbana-Champaign, Urbana, IL 61801

[#]Institute for Genomic Biology, University of Illinois at Urbana-Champaign, Urbana, IL 61801

Introduction: The stem cell niche provides extrinsic signals that regulate stem cell fate. The hematopoietic stem cell (HSC) niche in the bone marrow (BM) contains cells, the extracellular matrix (ECM), and ECM-bound or soluble biomolecules that present chemotactic, durotactic, and paratactic cues as well paracrine and juxtacrine signals to drive HSC fate. This niche regulates HSC fate decisions and hematopoiesis, the physiological process responsible for generating the body's blood and immune cells. While multiple cell types have been proposed as key regulators of HSC fate, the direct influence of heterotypic cell interactions via paracrine and juxtacrine signaling remain relatively unknown. We have developed a microfluidic platform to create microgels containing counter gradients of cell and matrix environments to determine differences in primary mHSC fate decisions in response to niche cell co-cultures using monolithic and counter-gradient 3D hydrogel environments designed to alter diffusion-mediated paracrine signaling.

Materials and Methods: Murine hematopoietic stem/progenitor cells (HSPCs) were isolated via FACS as $Lin^{-}ckit^{+}Sca1^{+}$ (LSK) or $LSKCD34^{-}$ BM cells. HSPCs were co-cultured with Lin^{+}BM cells under liquid as well as collagen, gelatin, and HA hydrogel cultures; hydrogel density as well as both HSC density and the HSPC:niche cell ratio (1:10, 1:100, 1:500) were varied to alter cell-cell communication. Cell-laden microgel constructs were polymerized thermally or via UV exposure and cultured over a period of 7 days with liquid co-culture used as a non-limited diffusion control. Hydrogel mechanical and diffusive properties were determined via compression and FRAP assays, respectively. HSPC proliferation was tracked via a fluorescent cell division assay while post culture HSPC fate was analyzed via FACS and gene expression analysis. A previously described microfluidic mixing device was used to create counter-gradient hydrogels comprised of HSPCs as well as multiple niche cells. Systematically varying HSPC:niche cell ratio as well as hydrogel type and density provides a rich landscape for examining the functional role of paracrine signaling on HSC fate.

Results and Discussion: Cell viability was not acutely affected by hydrogel gelation; higher HSPC viability and expansion (~50%) was observed for smaller gel volumes (50μL vs. 100μL). After 2 days of culture, FACS analysis indicates significant difference in HSPCs quiescence/maintenance (LSK) vs. myeloid lineage (CMP) differentiation due to both gel environment and niche cell co-culture. HSPCs showed increased specification towards myeloid lineages with increasing Lin+ niche cell co-culture. Significant differences were also observed between liquid and hydrogel conditions, with reduced niche cell effects observed in hydrogel environments that limited biomolecule diffusion. These results suggest a strong influence of niche cell interactions on HSPC fate.

Conclusions: We have established a biomaterial platform to explore the influence of niche cell mediated paracrine signaling on HSPC fate decisions. Ongoing experiments are examining longer culture periods and incorporating increasingly primitive niche cell populations (vascular endothelial cells, mMSCs) to alter the balance between lineage specification and quiescence. Counter gradients of quiescent and active niche populations are being incorporated into a single heterotypic microgel environment as an ectopic BM culture mimic to examine the dynamics of HSC fate decisions.

Acknowledgements: American Cancer Society, Illinois Division, Inc. (#160673, #189782).

(453g) Characterization of Acellular Matrix Derived From ESC Aggregates As Bioactive Scaffolds

Sébastien Sart, Teng Ma and Yan Li, Chemical and Biomedical Engineering, Florida State University, Tallahassee, FL

In recent years, embryonic stem cells (ESCs) have emerged as a powerful tool for the potential treatment of various incurable diseases. For therapeutic perspectives, mimicking stem cell niche or microenvironment, including *ad equate* cell-cell interactions and combination of extracellular matrix (ECM) molecules and biochemical factors, is essential for efficient pluripotent stem cell (PSC) expansion and differentiation. Acellular matrices derived from ESC aggregates, undifferentiated or embryoid bodies (EBs), can serve as potential scaffolds providing a more controlled bioactive three-dimensional microenvironment.

In this study, different chemical acellularization methods (i.e. Triton X-100, ammonia solution, SDS, and their combinations) are compared in association with DNAse treatment. The optimized protocol efficiently preserves ECM expression and improves DNA removal of ESC aggregates. These results highlight that the various defined structures generated from ESC aggregates could be efficiently acellularized and used as intact ECM-derived scaffolds. Next, ECM (fibronectin, laminin, collagen type IV, and vitronectin) expression is kinetically assessed for ESC monolayer, undifferentiated aggregates, and differentiated EBs with or without acellularization. Confocal microscopy and in situ ELISA are used to characterize ECM expression qualitatively and quantitatively. It is observed that ECM expression pattern is dynamically remodeled during expansion and differentiation. These results indicate that various culture conditions enable to generate different ESC-derived structures composed of specific ECM molecules composition. Finally, to prove the effectiveness of such scaffolds for ESC expansion and differentiation, undifferentiated ESCs are seeded on these ESC-ECM derived scaffolds and the markers of pluripotency and lineage differentiation are kinetically monitored to measure the degree of stemness and commitment of the newly seeded cells.

In summary, these results open the way towards the development of characterized bioactive scaffolds for more rapid and homogeneous ESC expansion and specific lineage differentiation.

(453h) Implantation of Vascular Grafts Made From Small Intestinal Sub-Mucosa and Hair Follicle Stem Cells in an Ovine Animal Model

Sindhu Row[1], Evan M. Schlaich[1], Hao-Fan Peng[1], Daniel D Swartz[2,3] and Stelios T. Andreadis[1,3,4], (1)Chemical and Biological Engineering, University at Buffalo, The State University of New York, Buffalo, NY, (2)Pediatrics, Women and Childrens Hospital of Buffalo, University at Buffalo, The State University of New York, Buffalo, NY, (3)Center of Excellence in Bioinformatics and Life Sciences, University at Buffalo, The State University of New York, Buffalo, NY, (4)Department of Biomedical Engineering, University at Buffalo, State University of New York, Buffalo, NY

Introduction: Cardiovascular disease is the leading cause of mortality worldwide. Regarded as the therapeutic gold standard, treatment by autologous vein grafting has several technical and patient-related risks. Alternatively, engineered grafts have shown to be one promising solution. Synthetic grafts have complications with compatibility, whereas natural, cell-seeded biomaterial scaffolds have been proven to perform better at the blood-material-interface. In this study, we aim to engineer fully functional and implantable arterial grafts made from human hair follicle derived smooth muscle cells (HF-SMC), ovine pulmonary artery endothelial cells, and small intestine submucosa (SIS).

Materials and Methods: Tubular tissues constructs were prepared from SIS and HF-SMC were incorporated in the vessel wall using fibrin glue. In order to promote living tissue development and improve the function of our TEVs, we mechanically stimulated the HF-SMCs within the construct using biomechanical forces and the stimulatory effects of TGFβ-1, insulin, and ascorbic acid. Using a customized bioreactor, arterial shear stress was achieved within our constructs. Prior to implantation, we investigated *in vivo* effects on parameters sustaining long-term patency of our grafts with the use of an arterio-venous shunt bypass model that we developed in our laboratory. Long-term patency of SIS grafts was evaluated when implanted as an interpositional graft in an ovine model, and followed for up to 12 weeks.

Results and Discussion: The grafts exhibited high burst pressure (1123327 mmHg, n=5) and similar compliance as compared to the native artery within physiological range (SIS=8.42+/-4.51, n=7; artery=8.62+/-2.99%, n=4, P>0.05, 0-120mmHg). Under dynamic strain, HF-SMC aligned circumferentially and showed collagen and elastin synthesis, as evidenced by histological examination. As expected, application of shear stress induced alignment of endothelial cells in the lumen in the direction of flow. The implanted tissues were monitored biweekly using Doppler ultrasound to document patency and measure blood flow rates through the grafts. At 4 and 12 weeks we also performed angiography to examine patency and measure blood flow rates through the grafts. After angiography the tissues were explanted and processed for histology and immunostaining. We found that all implants remained patent, contained a uniform endothelial monolayer and a robust vascular wall with numerous smooth muscle cells. Explanted grafts showed infiltration of host cells, robust ECM remodeling and vascular contractility similar to that of ovine arteries, demonstrating that our strategy was successful in engineering implantable arterial substitutes.

Conclusions: Our results demonstrate that SIS based vascular grafts with SMC from human HF-derived mesenchymal stem cells can be implanted in the arterial system of a large animal model. The patency of the grafts and their capacity for remodeling and development of vascular function suggest that this approach is very promising and may be of clinical significance.

Pete DeComo
American Institute of Chemical Engineering (AiChE)
October 31, 2012
Summary Abstract

Translational early stage Medical device development and the evolution of a technology and a company to the point of ultimate commercialization can be fraught with pitfalls. This presentation will highlight and illustrate real life challenges that occur during the product development process which must be overcome in order to obtain regulatory approvals followed by successful commercialization and a meaningful monetary exit for private and institutional investors.

The progress of a company is often complicated and costly due to the development of new and potentially disruptive technologies which are frequently the first of their kind in a targeted market and therapeutic application. Protracted regulatory approvals, product development challenges, scarcity of investment capital and human resources are just a few of the issues that can contribute to the well known phenomena of only one in ten start-up companies having significant success.

Early stage companies are often constrained by limited resources but nevertheless must successfully maneuver through these challenges in order to successfully capitalize the company and bring to market a revolutionary therapeutic modality that will improve the clinical and quality of life outcomes for thousands of patients, while at the same time reducing the total cost of care for the health care system.

(469b) Commercialization of Biomedical Technologies From University Research

Marc Malandro, Office of Technology Management, University of Pittsburgh, Pittsburgh, PA

Commercialization of technologies derived from university-based research must fit into the larger mission of the university. As such, technology transfer take many forms from dissemination of results in scientific publications to students working for companies to corporate sponsored- and focused-research to licensing of technologies developed at academic institutions. All of these activities are required to ensure that research results are making their way to the commercial marketplace for the benefit of society. The University of Pittsburgh, in close collaboration with its related health system the University of Pittsburgh Medical Center (UPMC) is committed to focusing research and commercial partnerships on areas of unmet clinical need. The process of commercialization at the University of Pittsburgh will be discussed, with particular focus on medical devices and the relationship of the engineering faculty and faculty in the schools of the health sciences. Case studies on commercialization will be presented, both licenses to established companies and the creation of spin-off companies. Finally, new approaches to facilitating collaborative research and translation in medical device development will be discussed.

(469c) Translational Research Initiatives within a University Setting

Pratap Khanwilkar[1], Marc Malandro[2], Mark Redfern[3], Stephen Badylak[4], Alan Hirschman[1] and Harvey Borovetz[5], (1)Bioengineering, University of Pittsburgh, Pittsburgh, PA, (2)Office of Technology Management, University of Pittsburgh, Pittsburgh, PA, (3)Swanson School of Engineering, University of Pittsburgh, Pittsburgh, PA, (4)Surgery, University of Pittsburgh, Pittsburgh, PA, (5)Bioengineering and Chemical Engineering, University of Pittsburgh, Pittsburgh, PA

Providing quality health care to patients that is accessible and affordable is a key goal for the activities of the nationally ranked Department of Bioengineering at the University of Pittsburgh. Traditionally, this has been done through basic and applied R&D by multidisciplinary teams of clinicians, bioengineers and other scientists.

Since 2011, the Swanson School of Engineering at the University of Pittsburgh has initiated two programs to help translate these innovative ideas and technologies developed by its R&D teams into clinical practice and commercialization.

In one program, the Center for Medical Innovation, a clinician with an idea for a health care improvement is paired up with an engineering faculty who can help realize the clinician's vision. Seed funding is provided. This is the 'bedside to bench' phase.

In a complementary program, called the Coulter Translational Research Partners II (TPII), the focus is on projects that have progressed substantially on the R&D front and need that 'last mile' of support, both financially and in business and commercialization acumen to successfully convert the technology into a start-up company or a license to an existing and established company. The medical technology developed must be a result of a partnership between a clinician and a bioengineering faculty at the University of Pittsburgh.

These 'Translational Research" programs require a global multi-disciplinary approach and teamwork both within the University of Pittsburgh's various schools, colleges, institutes and centers and a constructive engagement of the external business community.

The Katz Business School, the Law School through its Innovation Practice Institute and its Health Care Law program, the Schools of Health Sciences not only through the McGowan Institute but also through its Clinical Translational Science Institute, and the Offices of Technology Management and Enterprise Development are active and engaged partners in this effort. The Swanson School of Engineering is also developing a professional MS program to train student entrepreneurs interested in medical product innovation.

Essential partners outside of Pitt who have engaged to help achieve these program's goals are local and Silicon-Valley based venture capitalists, angel investment groups, corporations large and small, entrepreneurs, and content experts in areas such as regulatory affairs, reimbursement, marketing, finance, and intellectual property management.

Early indications point to these programs filling an unmet need within academia: for seed-stage CMI support, 32 proposals were received from a range of clinical areas and are being vetted to select 5 for further funding of about ~$10,000-$25,000 apiece in the May-June 2012 timeframe. A second round of solicitations for this program will begin in July, 2012. For the later-stage Coulter TPII program, 33 proposals were received from various clinical areas including cardiology, nursing, ophthalmology, radiology, oncology, and dentistry. Presently, 9 proposals are undergoing extensive commercialization due diligence which includes the development of a 15-minute business presentation and of a 5-page Executive Summary of the Business Plan for that particular technology. Of these 9, it is anticipated that 5 projects may be funded at about $100,000 per project for 1 year starting July 2012.

The Coulter Program is funded by the Wallace H. Coulter Foundation and the University of Pittsburgh for 5 years. This program has demonstrated its success at other prestigious Biomedical Engineering departments and Universities throughout the US and its results have been lauded by industry, investors and the federal government. In Phase I of this Program, about 7.5 times follow-on business funding has been received by projects funded by Coulter.

It is anticipated that innovative multidisciplinary programs such as these with measurable translational outcomes will lead to improved, high quality and cost-effective patient care, an imperative in this day and age, as well as contribute to regional economic development.

(483a) Active Stealth Signaling with a Synthetic 'self' Peptide

Dennis E. Discher, Chemical and Biomolecular Engineering, University of Pennsylvania, Philadelphia, PA

Foreign particles and cells are rapidly cleared from the body by professional phagocytes that incessantly encounter and recognize 'self' cells. The membrane protein CD47 is reportedly a 'Marker of Self' in mouse that impedes phagocytosis of self cells, with signaling through the highly polymorphic receptor CD172a on phagocytes. Here, minimal 'Self' peptides were computationally designed from $CD47^{human}$, synthesized with anchoring groups, and attached to virus-size nanoparticles for intravenous injection into NOD.SCID (NSG) mice that express a unique $CD172a^{NSG}$ (SIRPA) compatible with $CD47^{human}$. 'Self' peptides delay splenic macrophage clearance of particles with an exponential advantage in persistent circulation, which we use to enhance near-infrared imaging of human tumor xenografts by >10-fold. The affinity of $CD47^{human}$ for $CD172a^{NSG}$ proves weak but within the broad range (0.1 ~ 5 micro-Molar) measured for ten constructed variants of $CD172a^{human}$; several versions of 'Self' peptide are likewise shown to bind and potently inhibit nanoparticle uptake by an unexpected cytoskeletal mechanism. The reductionist approach reveals the importance and utility as well as some limits of a human 'Self' peptide.

(483b) Exploring Peptide-Based Nanostructures As Effective Drug Carriers

Ran Lin, Andrew G. Cheetham, Pengcheng Zhang and **Honggang Cui**, Chemical and Biomolecular Engineering, Johns Hopkins University, Baltimore, MD

Peptides can assemble into a variety of discrete nanostructures that could serve as biodegradable drug carriers for cancer chemotherapy. In this work, we demonstrate our rational design of short peptide conjugates to create various nano-objects such as spheres, cylinders, and cylinders with a multicompartment core. The critical micellization concentrations (CMC) of these conjugates can be tuned by molecular design, ranging in the nanomolar to micromolar scale. We evaluated the potential use of these nanostructures as drug carriers. In our drug encapsulation experiments, higher loading capacity is often achieved using peptides with relatively larger volume of the hydrophobic segments. In addition, by directly conjugating the anticancer drugs such as paclitaxel to a rationally designed short peptide, a quantitative and high drug loading can be achieved, and the resulting nanostructures can be used to encapsulate a different type of drugs. Our cytotoxicity experiments reveal that these supramolecular nanostructures can work effectively against a number of cancer cell lines including breast cancer, prostate cancer and ovarian cancer.

(483c) Dendron-Based Micelles: A Potential Nanocarrier Platform

Ryan Pearson, Jin Woo Bae, Hao-Jui Hsu, Sayam Uddin and **Seungpyo Hong**, Biopharmaceutical Sciences, University of Illinois at Chicago, Chicago, IL

Dendron-based amphiphilic copolymers possess unique features such as well-defined structure, self-assembling capability, and precise controllability over functionalities and high hydrophilicity. To assess the ability of dendron-based copolymers for use in drug delivery, we have synthesized novel, highly PEGylated dendron copolymers (PDC) with controlled hydrophilic-lipophilic ratios (HLs) ranging from 52:48 to 91:9. We then systematically compared their self-assembly properties with those of linear-block copolymer (LBC) counterparts at similar HLs. PDCs were synthesized via a multistep synthetic procedure. A hydroxyl-terminated generation 3 (G3) polyester dendron bearing a focal acetylene moiety was conjugated with poly(e-caprolactone) (PCL) by 'click' chemistry followed by PEGylation of the dendron surface with multiple methoxy poly(ethylene glycol) (mPEG) moieties. Four types of both PCL-G3-mPEG PDCs and PCL-mPEG LBCs with different block lengths were prepared and their structures were confirmed using ^{1}H NMR, FT-IR, and GPC. The critical micelle concentration (CMC) of PDCs was measured as low as 1.17×10^{-8}. The CMC values of the LBCs were up to 2 orders of magnitude higher than those of the PDCs at the same HLs, demonstrating superior thermodynamic stability of the PDC micelles. TEM images revealed that the PDCs formed spherical micelles (20-30 nm in diameter) with narrow size distributions, which was further supported by dynamic light scattering measurements. To understand in molecular detail the differences in self-assembly between PDCs and LBCs we utilized atomistic molecular dynamics (MD) simulations. Individual PDC copolymers were observed to adopt a stable conical shape in water, which is ideal for precise packing into spherical structures. Interestingly, the core was more completely covered by PEG in PDC micelles. The higher PEG surface coverage of PDC micelles may result in longer circulation half-lives due to decreased opsonization and enhanced accumulation at the therapeutic target site. Each micelle formulation was found to exhibit controlled release properties over 3 days using indomethacin as a model drug. PDCs exhibited no apparent toxicity up to concentrations as high as 100 mM evaluated by MTS assay. The low CMC at high HLs, controlled micelle morphologies, controlled release properties, and high PEG surface coverage all indicates the promising potential of the PDCs as a nanocarrier platform.

(483d) Microfluidic Cell Deformation As a Robust, Vector-Free Method for Cytosolic Delivery of Macromolecules

Armon Sharei[1], Janeta Zoldan[1], Andrea Adamo[2], Woo Young Sim[1], Nahyun Cho[1], Emily Jackson[1], Shirley Mao[1], Sabina Schneider[1], Abigail Lytton-Jean[1], Jungmin Lee[1], Daniel A. Heller[3], Robert S. Langer[4] and Klavs F. Jensen[5], (1)MIT, Cambridge, MA, (2)Chemical Engineering, Massachusetts Institute of Technology, Cambridge, MA, (3)Department of Chemical Engineering, Massachusetts Institute of Technology, Cambridge, MA, (4)Massachusetts Institute of Technology, Cambridge, MA, (5)Chemical Engineering, MIT, Cambridge, MA

Introduction

Intracellular delivery of macromolecules is a critical step in therapeutic and research applications. Nanoparticle mediated delivery of DNA and RNA, for example, is being explored for gene therapy[1], while protein delivery is a promising means of affecting cellular function in both clinical[2] and laboratory[3] settings. Other materials, such as small molecules, quantum dots, or gold nanoparticles, are of interest for cancer therapies[4], intracellular labeling[5], and single molecule tracking[6].

The cell membrane is largely impermeable to macromolecules. Many existing techniques utilize polymeric nanoparticles[7], or chemical modifications of the target molecule, such as cell penetrating peptides (CPPs) [8], to facilitate membrane poration or endocytotic delivery. In these methods, the delivery vehicle's efficacy is often dependent on the structure of the target molecule and the cell type. These methods are thus efficient in the delivery of structurally uniform materials, such as nucleic acids, but often ill-suited for the delivery of more structurally diverse materials, such as proteins[9] and some nanomaterials[10]. Moreover, the endosome escape mechanism that most of these methods rely on is often inefficient; hence, much material remains trapped in endosomal and lysosomal vesicles.

In this presentation, we detail a method for cytosolic delivery based on rapid mechanical deformation of the cell to produce transient membrane disruptions that facilitate the passive diffusion of material into the cell cytosol. This method was developed with the aim of delivering almost any macromolecule of interest to almost any cell type, at high throughput. Although shear based delivery methods have been demonstrated previously, they are unsuitable for some applications due to low viability and/or delivery efficiency[11]. The proposed technique, however, is distinct because it uses a physical constriction to deform and shear the cells in a controlled, reproducible manner. Unlike most current methods, this approach does not rely on electric fields, exogenous materials, endocytosis or chemical modification of the target molecule. Our data indicate that this method could be particularly advantageous for applications involving nanomaterials, proteins, or difficult-to-transfect cell types, such as immune cells and stem cells - all of which are often underserved by current methods.

Delivery method

We hypothesize that the rapid mechanical deformation of a cell, as it passes through a constriction with a minimum dimension smaller than the cell diameter, results in the formation of transient membrane disruptions or pores (**Fig. 1a**). The size and frequency of these pores would be a function of the shear and compressive forces experienced by the cell during its passage through the constriction. Material from the surrounding medium may then diffuse directly into the cell cytosol throughout the lifespan of these pores. Such an approach could theoretically enable the diffusive delivery of any macromolecule small enough to fit through the pores. To implement this approach, we generated a family of microfluidic devices with different constriction dimensions and numbers of constrictions in series.

Each device consists of 45 identical, parallel microfluidic channels, containing one or more constrictions, etched onto a silicon chip and sealed by a Pyrex layer. The width and length of each constriction range from 4-8um and 10-40um respectively. The current design is typically operated at a throughput rate of 20,000 cells/s, yielding close to one million treated cells per device prior to failure, due to clogging. The parallel channel design was chosen to increase throughput, while insuring uniform treatment of cells, because any clogging or defects in one channel cannot affect the flow speed in neighboring channels (the device is operated at constant pressure). Prior to use, the device is first connected to a steel interface that connects the inlet and outlet reservoirs to the silicon device. A mixture of cells and the desired delivery material is then placed into the inlet reservoir and Teflon tubing is attached at the inlet (not shown). A pressure regulator is then used to adjust the pressure at the inlet reservoir and drive the cells through the device. Treated cells are collected from the outlet reservoir.

An enabling platform

In order to illustrate our method's potential in addressing current delivery challenges, we conducted a number of proof-of-concept experiments using various nanomaterials and proteins of interest. Antibodies to tubulin, for example, were delivered (**Fig. 1b**) using this technique, yielding a diffuse distribution throughout the cell that would be consistent with cytosolic delivery. PEG1000 coated, 15nm gold nanoparticles were also delivered as measured by tunneling electron microscopy (TEM) of fixed cells (**Fig. 1c**). These nanoparticles appear to be mostly un-aggregated and are not visibly sequestered into endosomes. We were also able to verify the successful delivery of carbon nanotubes[12] (encapsulated by a DNA oligonucleotide) by flow cytometry and Raman spectroscopy. The aforementioned materials are currently difficult to deliver to the cell cytosol and each material often requires a specialized modification to facilitate delivery. In our work, all four materials were delivered to HeLa cells using the same set of conditions on a 10um-6umx5 device. In addition, this method has demonstrated the potential to accommodate a diverse range of cell types, including embryonic stem cells and primary lymphocytes (**Fig. 1d,e**), thereby signaling its potential for use across a range of research and perhaps clinical applications.

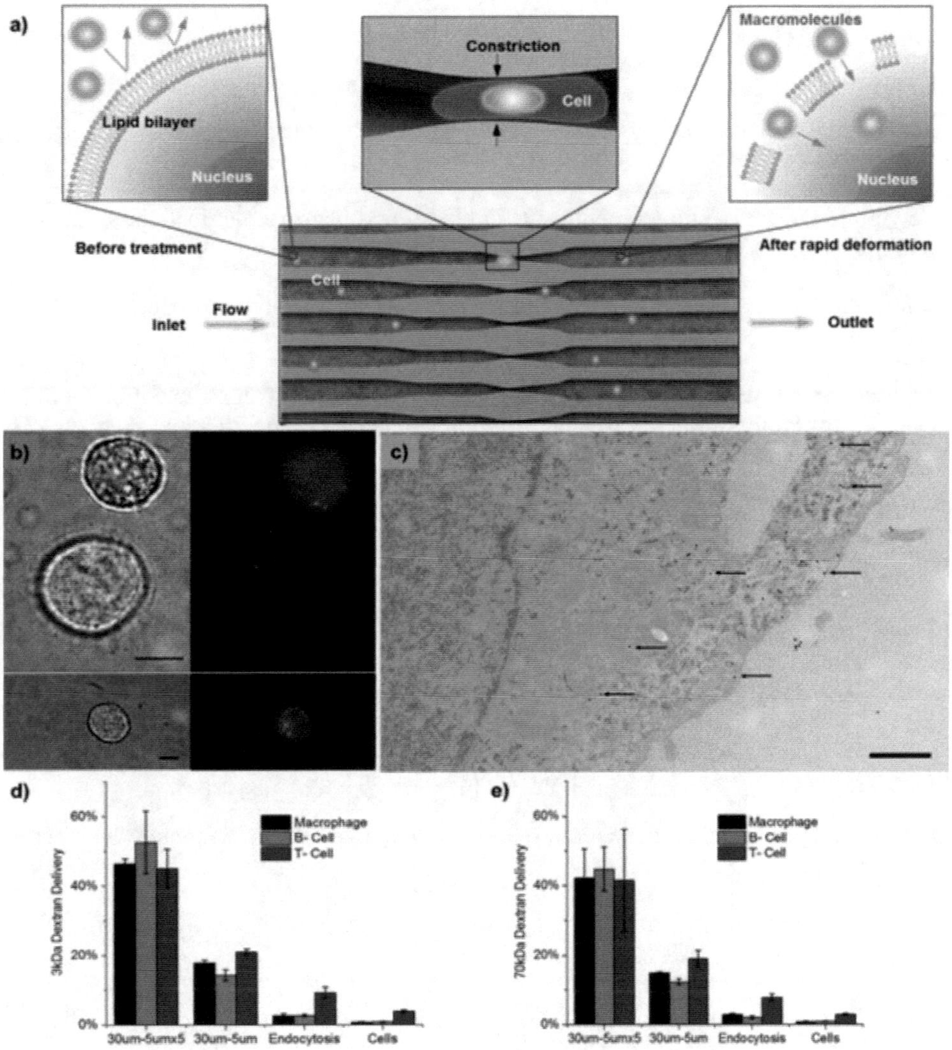

Figure 1. An overview of the delivery system and its capabilities. a) Illustration of delivery hypothesis whereby the rapid deformation of a cell, as it passes through a microfluidic constriction, generates transient membrane pores. Includes an electron micrograph of current parallel channel design with blue cells as an illustration. b) Fluorescent micrographs of HeLa cells immediately after delivery of antibodies to tubulin with an Alexa Fluor 488 tag. Scale bars at 5um. c) Tunneling electron microscopy (TEM) images of gold nanoparticles (some indicated by arrows) in a cell fixed ~1s after treatment. Scale bar at 500nm. d) Delivery of pacific blue labeled 3kDa dextran and e) fluorescein labeled 70kDa dextran to Macrophages (CD11b+), B lymphocytes (CD19+) and T lymphocytes (TCR+) derived from the blood of BL6 mice. The 30um-5umx5 and 30um-5um conditions denote two different device designs.

Summary

We have detailed a new method for cytosolic delivery that relies on the rapid mechanical deformation of a cell to induce transient membrane poration. This technique has demonstrated the potential to deliver a broad range of materials, some of which are challenging to use with current methods, to a variety of difficult-to-transfect cell types, including stem cells and immune cells. By providing such flexibility in application and obviating the need for exogenous materials or electrical fields, this method could potentially enable new avenues of disease research and treatment. Indeed our work has demonstrated this systems ability to deliver carbon nanotubes and antibodies to live cells – applications that could enable new sensing and imaging modalities.

References

1. Whitehead, K.A., Langer, R. & Anderson, D.G. Knocking down barriers: advances in siRNA delivery. *Nat Rev Drug Discov* **8**, 129-138 (2009).

2. Leader, B., Baca, Q.J. & Golan, D.E. Protein therapeutics: a summary and pharmacological classification. *Nat Rev Drug Discov* **7**, 21-39 (2008).

3. Kim, D. et al. Generation of Human Induced Pluripotent Stem Cells by Direct Delivery of Reprogramming Proteins. *Cell Stem Cell* **4**, 472-476 (2009).

4. Dhar, S., Daniel, W.L., Giljohann, D.A., Mirkin, C.A. & Lippard, S.J. Polyvalent Oligonucleotide Gold Nanoparticle Conjugates as Delivery Vehicles for Platinum(IV) Warheads. *Journal of the American Chemical Society* **131**, 14652-14653 (2009).

5. Michalet, X. et al. Quantum Dots for Live Cells, in Vivo Imaging, and Diagnostics. *Science* **307**, 538-544 (2005).

6. Dahan, M. et al. Diffusion Dynamics of Glycine Receptors Revealed by Single-Quantum Dot Tracking. *Science* **302**, 442-445 (2003).

7. Pack, D.W., Hoffman, A.S., Pun, S. & Stayton, P.S. Design and development of polymers for gene delivery. *Nat Rev Drug Discov* **4**, 581-593 (2005).

8. Heitz, F., Morris, M.C. & Divita, G. Twenty years of cell-penetrating peptides: from molecular mechanisms to therapeutics. *British Journal of Pharmacology* **157**, 195-206 (2009).

9. Yan, M. et al. A novel intracellular protein delivery platform based on single-protein nanocapsules. *Nat Nano* **5**, 48-53 (2010).

10. Derfus, A.M., Chan, W.C.W. & Bhatia, S.N. Intracellular Delivery of Quantum Dots for Live Cell Labeling and Organelle Tracking. *Advanced Materials* **16**, 961-966 (2004).

11. Hallow, D.M. et al. Shear-induced intracellular loading of cells with molecules by controlled microfluidics. *Biotechnology and Bioengineering* **99**, 846-854 (2008).

12. Heller, D.A., Baik, S., Eurell, T.E. & Strano, M.S. Single-Walled Carbon Nanotube Spectroscopy in Live Cells: Towards Long-Term Labels and Optical Sensors. *Advanced Materials* **17**, 2793-2799 (2005).

(483e) Self-Dispersing Drug Carriers for Pulmonary Delivery: Spreading of Aqueous Surfactant Solutions On Model Airway Surface Liquid Subphases

Amsul Khanal[1], Ramankur Sharma[2], Roomi Kalita[2], Fan Gao[3], Timothy Corcoran[4], Ellen Peterson[5], Todd M. Przybycien[6], Stephen Garoff[3] and **Robert D. Tilton**[6], (1)Department of Biomedical Engineering, Carnegie Mellon University, Pittsburgh, PA, (2)Department of Chemical Engineering, Carnegie Mellon University, Pittsburgh, PA, (3)Department of Physics, Carnegie Mellon University, Pittsburgh, PA, (4)Department of Medicine, University of Pittsburgh, Pittsburgh, PA, (5)Department of Mathematical Sciences, Carnegie Mellon University, Pittsburgh, PA, (6)Department of Chemical Engineering and Department of Biomedical Engineering, Carnegie Mellon University, Pittsburgh, PA

Persistent pulmonary infections are a leading cause of mortality in cystic fibrosis lung disease. This condition is associated with over-secretion of abnormally viscoelastic mucus that creates pulmonary airway obstructions. Current aerosolized antibiotic delivery techniques rely entirely on aerodynamic mechanisms to disperse and distribute these medications inside the lung after inhalation. The elements of obstruction common to cystic fibrosis and other lung diseases can decrease ventilation in portions of the lung, preventing aerosol drugs from reaching these zones. The unusual aerodynamics associated with obstructive disease can also cause highly non-uniform aerosol deposition patterns, further limiting distribution of drug throughout the lung. Poor distribution ultimately limits their efficacy. The inhaled antibiotics used to treat bacterial infections associated with cystic fibrosis lung disease often provide successful suppression of infection but rarely provide eradication. Drug resistance has also been associated with these therapies, likely due to the consistent delivery of sub-therapeutic doses at sites of infection. This group is investigating the potential to develop self-dispersing aerosol medications to improve the uniformity of drug delivery to obstructed lungs. An active agent is formulated with aqueous surfactant solutions that harness surface tension gradients to promote spreading on the airway surface liquid after aerosols deposit. Experiments on model systems show that surfactants significantly enhance spreading relative to saline controls as expected. Surfactant-laden aqueous drops when placed on entangled aqueous polymer solutions spread to form static lenses that persist on top of the subphase for tens of minutes, despite their complete miscibility. Our interests are focused on the relationship between formulation composition and the final spread area of deposited drops, a surrogate for the extent of drug dispersal after droplet deposition on the airway surface liquid. We use entangled aqueous poly(acrylamide) and porcine gastric mucin (PGM) solutions as subphases to mimic the physical properties of airway mucus plaques that are the main obstacles to uniform aerosol deposition in the lungs of patients with obstructive lung diseases. Solutions of the anionic surfactant sodium dodecyl sulfate (SDS), cationic surfactant cetyltrimethylammonium bromide (CTAB) or non-ionic surfactant tyloxapol contain dyes (absorptive or fluorescent) that serve as both model "drugs" and as tracers to measure the extent of drop spreading on the subphase. Regardless of the type of surfactant, when a drop of surfactant solution is placed on the polymer subphase, we observe that the dye spreads with the drop and that the surface tension of the subphase decreases in regions far outside the deposited drop, indicating that surfactant escapes across the drop contact line. This phenomenon is not observed with surfactant-free aqueous drops. The spread drop forms a finite, static lens that exists for on the order of 10 minutes while its contents slowly diffuse into the subphase. While Marangoni stresses initiate flow, a capillary force balance terminates the convective spreading process, resulting in the confinement of the spread solution to a well-defined static area. Spreading times are on the order of tens of seconds, considerably shorter than the timescales for diffusion into the subphase bulk or for subphase disentanglement. Spreading of aerosolized surfactant solutions is compared to the extent of spreading of macroscopic surfactant solution drops. The final spread areas for a macroscopic drop and the same deposited volume of aerosolized surfactant solution (aerosol size 2-10 um) are comparable, suggesting a similar spreading trend for aerosolized surfactant solutions and surfactant-laden drops on entangled polymer subphases that model the airway surface liquid.

(483f) Cathepsin B Degradable Peptidic Dendrimers for Drug Delivery

Rohit Kolhatkar, Ravi Shankar and Abhilash Samykutty, Biopharmaceutical Sciences, University of Illinois Chicago, Rockford, IL

INTRODUCTION: Polymers are widely used as drug delivery vehicles due to their ability to accumulate passively in tumor. Polymers can also be functionalized to achieve active targeting. Dendrimers offer several advantages over other conventional polymers. Here we report the use of peptidic dendrimers as drug delivery vehicles. These dendrimers are designed to degrade in the presence of an enzyme cathepsin B which is overexpressed at tumor site. A tetrapeptide *glycyl-phenylalanyl-leucyl-glycine* (GFLG) is used as a building block. The tetrapeptide GFLG is known to be recognized by cathepsin B. Two polymers with positive and negative surface charge were synthesized and characterized for their degradation profile in the presence of cathepsin B. By conjugation of a model drug 2-methoxyestradiol (2ME) to these polymers we show that peptidic dendrimers have the potential to be used as drug delivery vehicles.

EXPERIMENTAL METHODS: Positively and negatively charged peptidic dendrimers were synthesized using a convergent approach where tetrapeptide (GFLG) was added to the central core having four arms. Degradation profile of these polymers in the presence of enzyme cathepsin B and stability in the presence of PBS and acetate buffer was evaluated using HPLC. Glycine derivative of 2ME was attached to the tetrapeptide (GFLG) followed by conjugation of this unit (GFLGG2ME) to the central core to yield 2ME containing peptidic dendrimer. The activity of monomer and the polymer was evaluated by performing cytotoxicity studies using breast cancer cell line MDA-MB-231.

RESULTS AND DISCUSSION: Both positively and negatively charged peptidic dendrimers showed time dependant degradation in the presence of cathepsin B. Negatively charged dendrimer showed faster rate of degradation compared to positively charged amine terminated dendrimer. Polymers were stable in the presence of PBS and acetate buffer for 24 h. GI_{50} values (concentration to inhibit cellular growth by 50 %) for free drug 2ME, monomer GFLGG2ME were 0.89 μM and 7.6 μM respectively. Drug containing polymer (38 % wt/wt of drug) showed GI_{50} value of 8.5 μM while control polymer did not show any cytotoxicity. At equivalent drug concentrations, peptidic dendrimers were at least 8 times more active than *N*-(2-hydroxypropylmethacrylamide) (HPMA) copolymers containing 2ME (GI_{50} value of 114 μM).

CONCLUSION: Peptidic dendrimer has the potential to be used as carrier for 2-methoxyestradiol for breast cancer treatment.

ACKNOWLEDGEMENTS: This research was supported by Department of Biopharmaceutical Sciences, University of Illinois Chicago and Department of Defense (Breast Cancer Research) concept award (W81XWH-09-1-0687) to R. Kolhatkar.

(483g) Antitumor Efficacy Following the Intracellular and Interstitial Release of Liposomal Doxorubicin

Amey Bandekar, Biomedical Engineering and Chemical and Biochemical Engineering, Rutgers University, Piscataway, NJ and **Stavroula Sofou**, Biomedical Engineering, and Chemical and Biochemical Engineering, Rutgers University, Piscataway, NJ

pH-triggered lipid-membranes designed from biophysical principles are evaluated in the form of tar- geted liposomal doxorubicin with the aim to ultimately better control the growth of vascularized tumors. We compare the antitumor efficacy of anti-HER2/neu pH-triggered lipid vesicles encapsulating doxo- rubicin to the anti-HER2/neu form of an FDA approved liposomal doxorubicin of DSPC/cholesterol-based vesicles. The HER2/neu receptor is chosen due to its abundance in human breast cancers and its connection to low prognosis. On a subcutaneous murine BT474 xenograft model, superior control of tumor growth is demonstrated by targeted pH-triggered vesicles relative to targeted DSPC/cholesterol- based vesicles (35% vs. 19% decrease in tumor volume after 32 days upon initiation of treatment). Superior tumor control is also confirmed on SKBR3 subcutaneous xenografts of lower HER2/neu expression. The non-targeted form of pH-triggered vesicles encapsulating doxorubicin results also in better tumor control relative to the non-targeted DSPC/cholesterol-based vesicles (34% vs. 41% increase in tumor volume). Studies in BT474 multicellular spheroids suggest that the observed efficacy could be attributed to release of doxorubicin directly into the acidic tumor interstitium from pH-triggered vesicles extravasated into the tumor but not internalized by cancer cells. pH-triggered liposome carriers engi- neered from gel-phase bilayers that reversibly phase-separate with lowering pH, form transiently defective interfacial boundaries resulting in fast release of encapsulated doxorubicin. Our studies show that pH-triggered liposomes release encapsulated doxorubicin intracellularly and intratumorally, and may improve tumor control at the same or even lower administered doses relative to FDA approved liposomal chemotherapy.

(483h) Targeted Nitric Oxide Pretreatment Alters p53 and O6-Methylguanine-DNA Methyltransferase Activity Resulting in Enhanced Chemosensitivity in Glioma Cells

Shahana Safdar, Chemical and Biomolecular Engineering, Georgia Tech, Atlanta, GA, Courtney A. Payne, Spelman College, Atlanta, GA, Nam H. Tu, Georgia Institute of Technology, Atlanta, GA and Lakeshia Taite, Chemical and Biomolecular Engineering, Georgia Institute of Technology, Atlanta, GA

Introduction: Carmustine, 1,3-bis(2-chloroethyl)-1-nitrosourea (BCNU) and temozolomide (TMZ) are chemotherapeutics commonly used to treat glioblastoma multiforme. However a significant number of patients are resistant to these therapeutics resulting in poor patient survival. In the past, high doses of nitric oxide (NO) have been used to induce chemosensitivity in glioma cells. However the high doses coupled with the non-specific delivery of NO result in toxic effects to healthy tissue. Thus, to limit these side effects we have developed a targeted NO donor to enhance the efficacy of TMZ and BCNU on human malignant glioma cells.

Materials and Methods: To synthesize glioma-targeting NO donor, Chlorotoxin (CTX), a glioma targeting protein isolated from the venom of the Deathstalker scorpion, was reacted with NO gas for 24 hours in a hypoxic environment transforming CTX in to CTX-NO. Two malignant human glioblastoma cell lines T98 and U-87MG, normal human astrocytes (NHA) and human brain and microvascular endothelial cells (HBMEC) were used for *in vivo* studies. The cells were incubated for 2 hours with CTX-NO, after which they were incubated with BCNU (75 μM) or TMZ (50 μM) for 48 hours. At the end of the incubation period, cell viability, intercellular MGMT levels, or p53 expression was quantified.

Results and Discussion: TMZ and BCNU did not significantly change the viability of T98G cells. However, pretreatment with 2 μM of NO followed by TMZ or BCNU significantly decreased T98G cell viability to $76.8 \pm 8.5\%$ and $67.1 \pm 1.8\%$ of the control. Conversely, U-87MGs were more sensitive to both chemotherapeutics and NO pretreatment further enhanced this chemosensitivity (Fig. 1). CTX-NO resulted in no significant changes in the chemosensitivity of non-tumor control cells (Fig. 1). The resistance of tumor cells TMZ and BCNU is partially due to the presence of the DNA repair protein O^6-methylguanine-DNA methyltransferase (MGMT) therefore we investigated the effect of CTX-NO on MGMT levels in glioma cells. When T98G cells were treated with NO at concentrations of 1 μM or 2 μM, MGMT levels were decreased to $73.1 \pm 5.9\%$ and $60.1 \pm 9.9\%$ of the control. Another biomolecule that plays an important role in determining chemosensitivity is the protein p53 which a pivotal role in maintaining the balance between DNA repair mechanisms and induction of apoptosis. In T98G cells, which express mutant p53, CTX-NO treatment was able to significantly decrease in a dose dependant manner. Conversely, CTX-NO treatment did not affect p53 levels in U-87MG cells which express wild-type p53.

Conclusion: In this study we used small doses of a targeted NO donor, CTX-NO, to chemosensitize malignant glioma cells to chemotherapeutic agents, BCNU and TMZ. Furthermore the data showed that low levels of NO were able to decrease MGMT and p53 levels, both of which contribute to chemoresistance in glioma cells. Using targeted NO to sensitize cells towards chemotherapeutics holds great potential as part of a treatment plan for patients with various tumors that are normally unresponsive to chemotherapy.

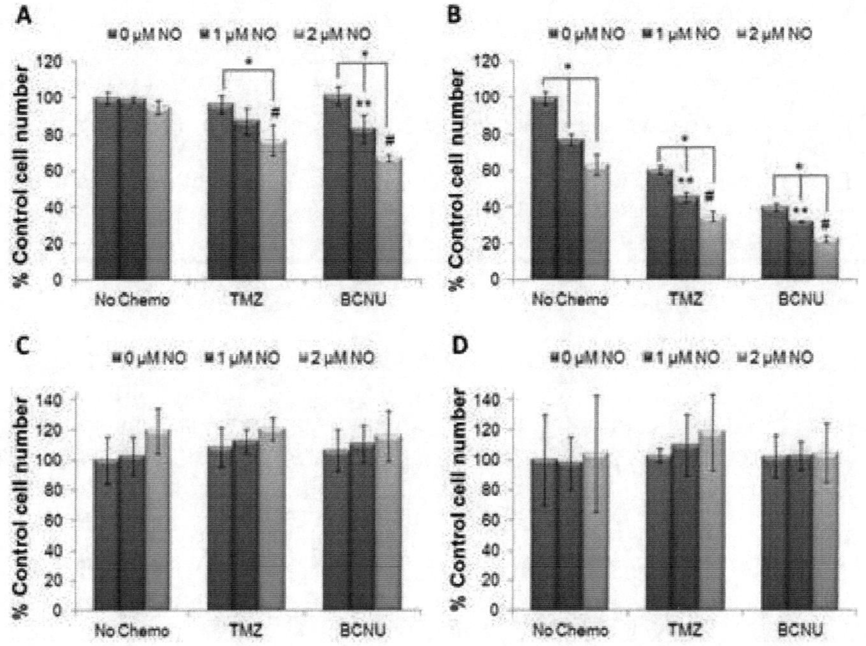

Figure 1: Cell viability of (A)T98G cells, (B)U-87MG, (C)NHA and (D)HBMEC. Cells counted are presented as a percentage of the number of cells that received no treatment. * compared to no pretreatment, ** compared to pretreatment with CTX-NO (1 μM), # compared to pretreatment with CTX-NO (2 μM) $p < 0.05$, n = 3

(483i) Strategies for Convection-Enhanced Drug Delivery

William L. Olbricht, School of Chemical and Biomolecular Engineering, Cornell University, Ithaca, NY

Convection-enhanced delivery (CED) is a novel method of delivering chemotherapeutics and other compounds to the brain for the treatment of primary brain tumors and certain neurological disorders. In CED, compounds are infused directly into the interstitium of brain tissue through a needle or cannula implanted in the brain. Because CED bypasses the blood-brain barrier, it can be used to deliver compounds that cannot be delivered to the brain by systemic means, including proteins, viral vectors, nucleotides, and molecules packaged in nanoparticles and liposomes.

The challenges in CED that we address are 1) achieving high infusion rates into the brain, and 2) controlling the spatial and temporal distribution of the infused material. High flow rates are important, because transport of the infused compounds near the needle is dominated by convection. For the drugs to penetrate far into the surrounding tissue, the convection rate must be sufficient to overcome the clearance of infused compounds due to a variety of elimination mechanisms.

To achieve high infusion rates into the brain, we have developed a series of microfabricated catheters containing microfluidic circuits that permit higher infusion rates and have several other advantages over needles currently used in CED. The microfabricated devices are much smaller than needles, they can be designed to deliver several therapeutic fluids in a programmed sequence, and they can be made flexible so that they can be implanted in the brain for extended times.

To control the spatial distribution of compounds once they have been infused into the brain, we expose regions of the brain tissue to safe levels of ultrasound. Ultrasound has been shown to increase mass transfer in tissue in several applications. By focusing the ultrasound on particular tissue volumes, the flow of infused fluid can be directed toward specific tissue targets identified from imaging studies. A series of experiments are described in brain phantoms and in anesthetized rodents to illustrate these CED strategies.

(489a) Multivalent Effectors to Control Stem Cell Differentiation

Anthony Conway and David V. Schaffer, Department of Chemical and Biomolecular Engineering, University of California, Berkeley, Berkeley, CA

Multivalent effectors to control stem cell differentiation

Numerous cellular signaling systems that regulate stem cell self-renewal and differentiation involve the assembly of multivalent ligands – oligomeric entities that present multiple binding sites and thereby bind multiple receptors on a target cell or stem cell – such as morphogens that naturally oligomerize, extracellular matrix protein engagement of integrins, and juxtacrine signaling between membrane-associated ligands and receptors. The resulting multivalent binding can be more potent than corresponding monovalent interactions, potentially through initiating a process of cellular receptor clustering that may promote enhanced signal transduction. However, the ability to control and monitor these naturally occurring multivalent interactions within a living cell is currently limited, since it is difficult to modulate the valency of naturally-occurring ligands, especially membrane-associated proteins. We have developed a means to synthetically conjugate the binding domains of normally cell membrane-associated ligands to the soluble biopolymer hyaluronic acid (HA). Specifically, using EDC/NHS chemistry, we functionalized the high molecular weight HA backbone and attached recombinantly produced ephrin-B ligand ectodomains via an added terminal cysteine. To precisely estimate the valency and molecular weight of the bioconjugates, we characterized the molecules using size exclusion chromatography and multi-angle light scattering.

To determine the effects of bioconjugate multivalency on the ability to cluster cell surface receptors, we incubated adult hippocampal neural progenitor cells (AHNPCs) with the multivalent molecule *in vitro.* Multivalent ephrin-B showed enhanced clustering of cognate Eph receptors as compared to antibody-clustered ligands. Next, to determine if increased receptor clustering could affect stem cell differentiation, cultures of AHNPCs were incubated solely with either antibody-clustered ligands or high conjugation ratio conjugates for six days. A maximum three-fold increase over the antibody-clustered ligand for the highest ratio conjugate was observed. To determine if the effect of multivalency in directing neuronal stem cell fate could be applied more generally to other cell types, cultures of human embryonic and induced pluripotent stem cells were incubated with the factors for 2-4 weeks. In both cultures a 2-fold increase in the fraction of neurons formed and a 4-fold increase in the fraction of dopaminergic neurons compared to antibody-clustered controls was observed. Cultures also showed increases in the midbrain-specific marker En1 and had significantly higher levels of dopamine. Finally, in an effort to validate the enhanced activity of this multivalent construct *in vivo,* we injected it directly into the brain of adult rats via intracranial stereotaxic injection. Brains injected with bioconjugates showed a three-fold increased ability to induce neuronal differentiation as compared to the antibody-clustered control, thus indicating that highly multivalent ligands potently enhance the fraction of new neurons formed compared to using ligands clustered by conventional means. These results have a variety of biomedical implications, in that they establish a general platform for creating highly bioactive, defined, and reproducible forms of protein-based ligands, which have the ability to more potently activate downstream effectors for use in basic research and therapeutic applications.

(489b) Engineered Microenvironment for Osteogenic Differentiation of Stem Cells

Ameya Phadke, University of California, San Diego, La Jolla, CA, Yu-Ru Shih, Bioengineering, University of California, San Diego, La Jolla, CA and **Shyni Varghese**, Bioengineering, University Of California, San Diego, La Jolla, CA

Bone is one of the most common transplanted tissues, second only to blood. Currently, autografts and allografts are the primary choice for treating compromised bone tissue, but both suffer from drawbacks such as donor site morbidity, scarcity, immunorejection, and disease transmission. In contrast, synthetic bone grafts with intrinsic osteoinductivity and osteoconductivity could provide an easy-to-manufacture and cost-effective therapeutic strategy for treating bone defects. However, employing biomaterial-based bone grafts to accelerate regeneration of critical bone defects requires a detailed understanding of bone extracellular matrix (ECM), its role in the differentiation of stem cells into osteoblasts, and tissue homeostasis. Recently, we have developed novel, synthetic bone-like ECM structures using principles of biomineralization. Thus-formed biomineralized matrices recapitulate various attributes of native bone ECM. Our ongoing studies involving human mesenchymal stem cells indicate that the biomimetic matrix directed their osteogenic differentiation both in vitro and in vivo. Furthermore, we have employed the biomineralized matrices as an "artificial niche" to understand the mechanism by which mineralized matrices induces osteogenic differentiation commitment of stem cells.

(489c) Variation of Oxygen in a Controlled Manner Markedly Enhances Multi-Stage Differentiation of Embryonic Stem Cells to Insulin Producing Cells

Amanda R. DiIenno, Jeffrey R. Millman, Anna Kokensparger and Clark K. Colton, Chemical Engineering, Massachusetts Institute of Technology, Cambridge, MA

Pluripotent stem cells (PSC) hold promise for cell replacement therapy, but efficient differentiation to desired cell types remains a major obstacle. Although most PSC research is performed in high, non-physiological O_2 environments, cells during embryonic development are actually exposed to much lower O_2. Here we report that variation of O_2 in a controlled manner markedly enhances multi-stage differentiation of embryonic stem cells to insulin producing cells. We differentiated CyT49 human embryonic stem cells (hESC), using a modification of the 5-stage protocol published by ViaCyte, Inc (D'Amour et al., Nat Biotech 2006, Kroon et al., Nat Biotech 2008) under different sequences of well-characterized pO_2 environments. We controlled cellular oxygen exposure through adhesion culture on highly O_2-permeable silicone rubber membranes so that the cells in contact with the membrane were exposed to the same pO_2 as in the gas phase.

Our strategy in selecting O_2 patterns to investigate was based on two hypotheses: (1) cells in the developing embryo are initially exposed to low O_2 levels that increase as the tissues become vascularized. (2) The effects of low O_2 on the early stages of differentiation can be manifested in later stages of differentiation. At each stage there were two steps. We started with stage 1at six different O_2 levels (1%, 3%, 5%, 8%, 10%, & 20%). Samples at the end of stage 1 were analyzed with qPCR and flow cytometry for markers of definitive endoderm (DE). In the second step, we carried the differentiation to completion. The one to three best oxygen levels that produced the highest fraction of DE cells were then used in stage 1, and the remaining stages were carried out at 20% O_2. Each set of complete differentiation experiments employed a control condition of constant 20% O_2 for comparison. The best stage 1 O_2 level was selected on the basis of the fraction c-peptide+ and/or insulin+ cells in the stage 5 preparation as well as the extent of gene expression of relevant pancreatic endoderm (PE) and beta cell markers, and this O_2 level was used for all subsequent experiments. This two-step process was repeated sequentially with stage 2, 3, and 4; the condition for stage 5 was fixed at 20% O_2. Beginning with stage 3 an additional O_2 condition of 40% was added to the conditions tested. All together, 12 different sequences of O_2 levels in successive stages were evaluated.

The fraction of cells containing insulin or c-peptide was similar. Results for 9 of the 12 sequences are summarized in Table 1 in terms of the pooled mean of the fractions of c-peptide+ and insulin+ cells.

Table 1: Fraction of cells that were c-peptide+ and/or insulin+ (Mean ± SD)

Conditions*	C-peptide+/ Insulin+ (%)
20-20-20-20	4 ± 1
5-20-20-20	9 ± 4
5-5-20-20	8 ± 2
5-5-5-(1 to 3)	8 ± 7
5-5-5-(5 to 40)	34 ±13

* percent O_2 in sequential stages

With normoxic conditions in all of the stages (designated 20-20-20-20), the fraction was about 4%. Reduction of O_2 to 5% in stage 1caused the fraction to double. A similar additional reduction of O_2 in stage 2 had no further effect. The effect of another sequential reduction to 5% in stage 3 depended on the oxygen level in stage 4. If O_2 was further reduced below 5% in stage 4, there was no change. However, in the range of 5 to 40% O_2 for stage 4, the fraction of c-peptide+/insulin+ cells jumped by a factor of four to a mean of 34%. At two conditions, the fraction was about 50%, but there was not a monotonic trend with O_2 level.

These results demonstrated that modulation of the O_2 level to which cells are exposed can have a pronounced effect on the outcome of directed differentiation protocols. In the present example, culture at reduced oxygen for the first three stages, followed by an increase in O_2 levels for the last two stages, produced an eight-fold increase in the fraction of c-peptide+/insulin+ cells.

(489d) Mesenchymal Stem Cell Therapeutics for Protection and Repair of Injured Tissues & Vital Organs

Martin L. Yarmush, Biomedical Engineering, Rutgers University and Massachusetts General Hospital, Piscataway, NJ

Recently there has been a paradigm shift in what is considered to be the therapeutic promise of mesenchymal stem cells (MSCs) in diseases of vital organs. Originally, research focused on MSCs as a source of regenerative cells by differentiation of transplanted cells into lost cell types. It is now clear that trophic modulation of inflammation, cell death, fibrosis, and tissue repair are the primary mechanisms of MSC therapy. This has been clarified in studies where delivery of growth factors, cytokines, and other signaling molecules secreted by MSCs is often sufficient to obtain the therapeutic effects. In this presentation, we provide a summary of the current knowledge on trophic mechanisms of MSC therapy in disease models of vital organs using models of acute liver failure, acute kidney failure, and spinal cord injury. Important issues regarding the optimal delivery methods of MSC therapy are discussed and critical gaps in our knowledge hampering experimental progress and clinical implementation are discussed.

(489e) Endothelial Cells Mediate Maturation of Human Embryonic Stem Cell Dervied Pancreatic Progenitors Into Insulin Expressing Cells

Maria Jaramillo[1], Saik Kia Goh[1] and Ipsita Banerjee[2], (1)Bioengineering, University of Pittsburgh, Pittsburgh, PA, (2)Chemical and Petroleum Engineering, University of Pittsburgh, Pittsburgh, PA

Diabetes and its complications affect 25.8 million people in the US and is a growing healthcare problem worldwide. While transplantation of pancreatic islets offer potential therapy, it is restricted by scarcity of donor islets. Effective use of hESC could overcome this limitation, however current approaches to differentiating hESCs to mature islets have been restricted in yield and functionality of the differentiated phenotype. In our work we propose to mimic *in-vivo* development during maturation of hESC derived pancreatic progenitor cells. Over past years developmental studies have established the critical involvement of endothelial cells in maturation and functionality of pancreatic islets, which is not surprising given the islet's dense vasculature.

We have developed a multi-stage protocol for directed differentiation of hESCs involving definitive endoderm induction, followed by pancreatic progenitor specification and final maturation into insulin producing cells. For the last step, we propose to co-culture the hESC-derived pancreatic progenitor cells with endothelial cells in accordance with developmental studies. We used two different cell types in our studies: Rat Heart Microvascular Endothelial Cells (RHMVEC) and Human Umbilical Vein Endothelial Cells (HUVEC). As a control experiment we co-cultured the pancreatic progenitor cells with NIH3T3 fibroblast to verify sensitivity to cell types. All direct contact co-cultures were FACS sorted and analyzed for gene and protein expression levels. Two different non-contact co-culture configurations were also investigated: transwell co-culture and endothelial cell conditioned media. Results were evaluated with respect to conventional maturation using notch inhibition by DAPT.

We evaluated each stage of differentiation using qRT-PCR, immunostaining and flow cytometry. Our protocol resulted in 89% Foxa2 positive endoderm and high, uniform nuclear staining for Pdx-1 at pancreatic progenitor stage with roughly 70% of the cells being Pdx1 positive as observed by flow cytometry. Maturation with DAPT resulted in 17,500 fold upregulation of insulin mRNA with 8% commitment, while parallel RHMVEC co-culture resulted in 20,000 fold upregulation of insulin mRNA with 34% commitment. HUVEC co-culture resulted in an even higher insulin expression, 7 times higher than RHMVEC. Control cultures with fibroblast did not show appreciable insulin upregulation, indicating the effect to be specific to endothelial cells. Alternative co-culture configurations also show promising results, with transwell dominating over conditioned media, but still lower than direct contact configuration.

Our results establish an alternate strategy for hESC maturation into islet cell types. Current investigation is underway to elucidate the mechanism of endothelial cell mediated maturation.

(489f) Human Induced Pluripotent Stem Cells Differentiate Into Contractile Vascular Smooth Muscle Fate Via Mesenchymal Stem Cell Intermediates: Implication for Cardiovascular Regeneration

Vivek K. Bajpai, Chemical & Biological Engineering, State University of New York at Buffalo, Amherst, NY and Stelios T. Andreadis, Department of Chemical and Biological Engineering, University at Buffalo, the State University of New York, Amherst, NY

Smooth muscle cells (SMC) maintain vascular homeostasis and play prominent role in vascular diseases. Induced pluripotent stem cells (iPSC) can provide an untapped resource of functional SMC for autologous cell based therapies and for creating models for studying vascular diseases. This would require efficient differentiation of iPSC towards SMC fate. However existing protocols of pluripotent stem cell differentiation are inefficient and primarily give rise to terminally differentiated SMC that critically limits their in-vitro expansion that is unequivocally desirable for regenerative medicine applications. To address this challenge, we developed a robust, stage-wise monolayer protocol of human (h)iPSC differentiation towards functional SMC fate. Employing this protocol, we demonstrate that hiPSC differentiate into functional smooth muscle cell fate via intermediate mesenchymal stem cell (MSC) stage where they are highly proliferative and amenable to differentiate along all mesenchymal cell types including osteogenic, chondrogenic and adipogenic lineages. Thus using this protocol, MSC can be kept in a proliferative state for 45 population doublings and tuned into contractile SMC with short-term induction treatment when required. In the first stage of differentiation hiPSC lines (generated from two different set of transcription factors) were induced to undergo epithelial to mesenchymal transition (EMT) by culturing them on matrigel-coated dishes in presence of EMT medium (MEF conditioned medium human embryonic stem cell medium supplemented plus FGF2). EMT was followed in real time using smooth muscle alpha actin (ACTA2) promoter driven ZsGreen lentiviral reporter. To ensure that ACTA2-ZsGreen reporter reliably captures hiPSC undergoing EMT, we derived stable ACTA2-ZsGreen lentiviral reporter hiPSC clones and by immunocytochemistry clearly demonstrated that undifferentiated OCT4+ colonies also lacked ACTA2-ZsGreen expression. However, as early as day 3 after induction of EMT cells in the periphery of hiPSC colonies assumed fibroblastic morphology and expressed ZsGreen, suggesting upregulation of ACTA2 promoter and EMT. Indeed, EMT was confirmed by RT-PCR for EMT markers like SNAIL, TWIST, SLUG, FOXC2, FSP1, CDH1, CDH2, fibronectin and vimentin. EMT cells (ZsGreen+) were enriched (stage 2) into MSC fate by culture on gelatin-coated surfaces in the presence of enrichment medium (M231 supplemented with 5%FBS, bFGF, EGF, insulin and heparin). During the enrichment stage, the fraction of ZsGreen+ cells was significantly increased for the both hiPSC lines tested as quantified by flow cytometry. Enriched cells displayed remarkable similarity with bone marrow derived MSC as they displayed MSC surface immunophenotype (CD105+/CD73+/CD44+/CD49b+/CD90+/CD45-/CD34-) and could be coaxed to differentiate into adipogenic, osteogenic and chondrogenic lineages as shown by oil red O, Von Kossa and alcian blue stains, respectively. Enriched cells were highly proliferative cells as they were kept in culture for at least 45 population doublings without reaching cellular senescence. While enriched cells were highly proliferative, they were weakly contractile, perhaps due to immature contractile machinery as evidenced by immunostaining for ACTA2, CNN1 and MYH11. To induce contractile phenotype (stage 3), enriched MSC were further treated with maturation medium (M231 plus TGF-β1 and heparin). Immunostaining of mature cells showed markedly increased filamentous organization of ACTA2, CNN1 and MYH11 proteins. Contractile phenotype was further confirmed by excellent contractility in response to receptor and non-receptor mediated vasoagonists exhibited by cylindrical tissue engineered vessels (TEV) made from mature cells.

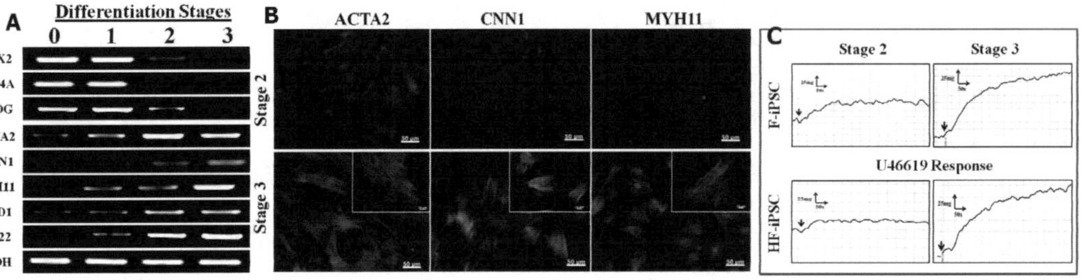

Figure 1. RT-PCR analysis of SMC specific and pluripotency markers during differentiation (A) Immunocytochemistry for SMC specific markers show filamentous organization of these contractile proteins after maturation treatment (B) Vasoconstriction of TEV made of enriched and mature cells for two hiPSC lines (C).

In summary, this study demonstrate that hiPSC undergo functional SMC fate through an intermediate stage of multipotent MSC. The iPSC derived SMC may be useful for studying the pathology of vascular diseases especially in context of SMC phenotypic conversion between synthetic and contractile fates and engineering SMC-containing tissues such as vascular grafts.

(550a) Drug Encapsulated Polymeric Microspheres for Temporally-Staged, Localized Brain Tumor Therapy

J. Alaina Floyd[1], Anna Galperin[2], Rohan Ramakrishna[3], Robert Rostomily[3] and Buddy Ratner[4], (1)Chemical Engineering, University of Washington, Seattle, WA, (2)Bioengineering, University of Washington, Seattle, WA, (3)Neurological Surgery, University of Washington, Seattle, WA, (4)Chemical and Bioengineering, University of Washington, Seattle, WA

The American Cancer Society 2012 estimate is that 22,910 malignant tumors of the brain or spinal cord will be diagnosed in the United Sates. Approximately 59.8% will be fatal. Despite current treatments, survival remains low with a median survival of approximately one year despite maximal therapy. Complete tumor removal cannot always be achieved because of the potential for brain damage, leading to cancerous regrowth and ultimately, patient death. In addition, these tumors can develop drug resistance, rendering chemotherapy ineffectual. We believe that the development of a topical, slow release, multi-drug delivery system (M-DDS) that is applied post-surgically directly in the tumor cavity could significantly increase life expectancy. Our M-DDS consists of multi-drug loaded poly(lactic acid) (PLA), poly(lactic-co-glycolic acid) (PLGA), and poly(ε-caprolactone) (PCL) microspheres suspended in an aqueous solution of a degradable, thermoresponsive poly(*N*-isopropylacrylamide) (PNIPAM) synthesized in our laboratory. At room temperature, PNIPAM suspends drug encapsulated microspheres that can be directly "sprayed on" the post-surgical site. The 37°C temperature of the brain would pass PNIPAM through its lower critical solution temperature, causing the polymer matrix to solidify and form an adherent gel layer with entrapped drug loaded microspheres on the surface of the brain. This would provide intimate contact with the remaining tumor cells, conforming to the convoluted morphology of the brain surface. Over time PNIPAM would degrade along with PLGA, PLA, and then PCL microspheres, releasing multiple chemotherapeutics at different rates directly to the tumor remnants to inhibit cancerous re-growth. This work focuses on microsphere characterization, model drug (rhodamine B (RB)) and chemotherapeutic encapsulation, and RB release from microspheres and microspheres entrapped in PNIPAM.

Both double and single emulsions have been applied to form PLA, PCL, and PLGA microspheres. Smooth, nonporous microspheres were produced with an adjustable size dependent on impeller speed during the emulsification process. RB, a hydrophilic dye and the model drug, was successfully encapsulated within PLA, PCL, and PLGA based microspheres and microsphere morphology remained unaffected. The encapsulation efficiency was polymer dependent; PLGA was most efficient at RB incorporation ($84.6 \pm 4.8\%$) due to its hydrophilic nature. An *in vitro* release from PLGA microspheres alone and PLGA microspheres entrapped in degradable PNIPAM was characterized in PBS at 37 °C. A biphasic release profile was seen for both conditions, with an initial, moderate, continuous release (passive diffusion) followed by a significant increase in the release curve slope (polymer degradation) before plateauing. Microspheres entrapped in PNIPAM had a two day delay of release compared to microspheres alone, with the release plateau occurring at day 22 versus day 20. In both instances, smaller microspheres (22.5 ± 7.8 μm) released faster than the larger microspheres (42.2 ± 15.4 μm) during the diffusion period due to a higher surface area to volume ratio, showing release profile tunability.

Our M-DDS has also been tested to visualize local and topical delivery on brain tissue. RB encapsulated PLGA microspheres suspended in degradable PNIPAM were "sprayed on" a rat brain that was preserved at 37 °C. The PNIPAM solution underwent a phase transition on the rat brain, showing stable adhesion to the tissue and efficient entrapment of drug loaded microspheres to a localized site (Figure 1).

The chemotherapeutic gefitinib has also been successfully encapsulated in PLGA microspheres (Figure 2). These gefitinib loaded PLGA microspheres will be tested against a C6 glioma cell line to explore drug efficacy along with PCL microspheres encapsulating the chemotherapeutic temozolomide. A pilot animal study will be conducted to test the effect of the delivery system on a rat model and the diffusion of the chemotherapeutics through the brain. Our M-DDS has the potential to provide a localized, tunable time release of multiple chemotherapeutics that will directly impact tumor remnants to inhibit cancerous re-growth, overcome tumor drug resistance, and increase patient life expectancy.

a) b)

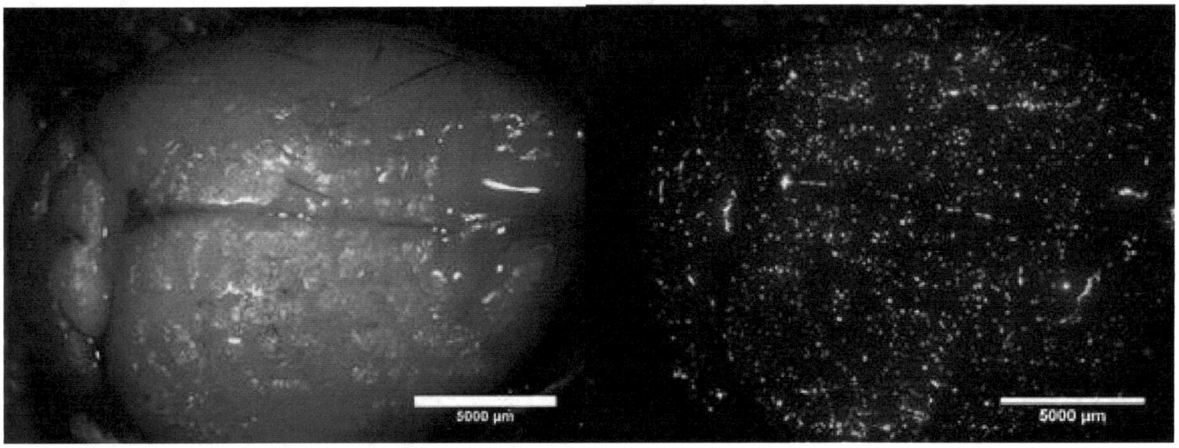

Figure 1: Sprayed RB encapsulated PLGA microspheres entrapped in degradable PNIPAM on a rat brain at 37 °C viewed with a) bright field and b) fluorescent microscopy

a) b)

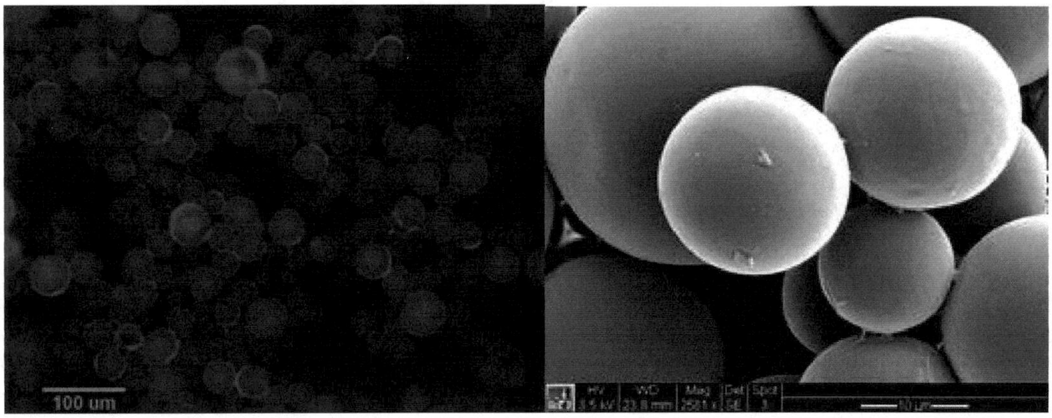

Figure 2: Gefitinib encapsulated PLGA microspheres viewed by a) fluorescent and b) scanning electron microscopy

(550b) Redox Responsive Polymeric Nanocapsules for Protein Delivery

Muxun Zhao, Chemical and Biomolecular Engineering, UCLA, Los Angeles, CA and Yi Tang, Chemical and Biomolecular Engineering, University of California, Los Angeles, Los Angeles, CA

Virtually all human cancer cells have elaborate anti-apoptotic strategies to overcome apoptosis, which is a vital anti-neoplastic mechanism to obstruct tumor progression. Therefore, the ability to rapidly resurrect the apoptosis circuitry in tumor cells can be a highly effective option for cancer treatment. Although different strategies have been developed to counter the individual inactivation mechanisms, a more potent chemotherapy option is to directly arm the cancer cells with executioner proteins or apoptosis-inducing proteins that are not targeted by these aforementioned anti-apoptotic maneuvers. From a therapeutic perspective, protein-based approaches are safer than gene therapy because no random or permanent genetic changes are involved, and only transient actions of proteins are needed for the desired results. However, compared to small molecule therapeutics, proteins suffer from serum instability and are inefficient or unable to penetrate the cell membrane. As a result, protein-based antitumor options still remain elusive, despite its numerous therapeutic targets and vast potential. We have developed a simple and effective method of delivering proteins intracellularly. Using this method, the avian virus-derived protein Apoptin, which is a selective killer of various tumor cells, have been delivered to cancer cells in cell culture and in xenograft mouse tumors.

Apoptin is a small protein (121 amino acids) isolated from chicken anemia virus (CAV) that induces p53-independent apoptosis in a tumor-specific way. In a variety of tumor cell lines, Apoptin becomes phosphorylated, enters the nucleus, and induces apoptosis. In sharp contrast, Apoptin is unphosphorylated in normal cells and stays in the cytoplasm. An important feature of Apoptin is that it can recognize early stages of oncogenesis and it can induce apoptosis. Currently known Apoptin targets include DEDAF, Nur77, Nmi, and Hippi, some of which are p53-independent signaling proteins in the apoptotic pathway. Due to its high selectivity and potency, Apoptin has become an attractive antitumor target for gene therapy approaches.

To deliver apoptin protein directly, we first encapsulated recombinant maltose-binding-prortien (MBP) apoptin fusion protein in a degradable polymeric nanocapsule. The nanocapsule is formed by first electrostatically depositing positively charged monomers on to the surface of the MBP-apoptin complex (MBP-apoptin forms a 30-40 mer multimeric complex with an estimated size of 35 nm). Followed by addition of a redox-sensitive crosslinker, polymerization is initiated to form a thin, positively charged polymer layer that encapsulates the MBP-apoptin. The particles are found to be uniform in size and charge. When added to cancer cell cultures, the nanocapsule readily penetrated the cell membranes, disintegrated to release the MBP-apoptin and induced rapid apoptosis of the cells. In contrast, the nanocapsule had no apoptotic effect on primary cell lines such as HFF, highlighting the differential targeting on tumor cells only. Confocal microscopy of rhodamine labeled MBP-Apoptin showed that the protein entered the nucleus of cancer cells only, while remained in cytosol of HFF. Additional controls using nondegradable nanocapsules and apoptin mutants were performed to confirm the proposed mechanism of action of the nanocapsulated MBP-apoptin. When injected intratumorally into mice carrying xenografted MCF7 tumor. The nanocapsule significantly inhibited the progression of tumor growth in comparison to the controls, demonstrating the potential utility of this polymer-protein complex as a cancer therapeutic.

(550c) Stabilization of Pneumoccoccal Surface Protein A in Polyanhydride Nanoparticles: Consequences for the Design of a Pneumonia Vaccine

Shannon Haughney[1], Latrisha Petersen[1], Janice King[2], Amanda Ramer-Tait[3], Amy Schoofs[3], David Briles[2], Michael J. Wannemuehler[3] and Balaji Narasimhan[4], (1)Chemical and Biological Engineering, Iowa State University, Ames, IA, (2)Department of Microbiology, University of Alabama Birmingham, Birmingham, AL, (3)Department of Veterinary Microbiology and Preventive Medicine, Iowa State University, Ames, IA, (4)Department of Chemical and Biological Engineering, Iowa State University, Ames, IA

Pneumoccoccal surface protein A (PspA) is a choline-binding protein found on the surface of all pneumococcal strains and has been shown to be critical to the virulence of infections with *Streptococcus pneumoniae*. PspA plays two different roles in invasive infection and nasopharyngeal carriage. In invasive, systemic infections with *S. pneumoniae*, PspA prevents the deposition of complement, an immune protein that assists antibodies and phagocytic cells. PspA also plays a role in carriage by binding the bactericidal glycoprotein, apolactoferrin. Vaccination with PspA has been shown to be protective against a lethal challenge with *S. pneumoniae* making it a promising candidate for use in vaccines. Nanoparticles and microparticles made from polyanhydrides have been shown to successfully preserve antigenicity of proteins throughout the manufacture, storage, and release steps. These polymers have excellent biocompatibility and have been shown to have non-toxic, non-mutagenic degradation products. Another important benefit that leads to increased stabilization of protein is the use of a non-aqueous particle fabrication process, which not only prevents protein degradation during exposure to water, but also the loss of encapsulation efficiency observed in aqueous fabrication processes. Amphiphilic polyanhydrides are ideal candidates for vaccine delivery because of their ability to: stabilize and protect fragile antigens; provide a sustained release of antigen; and modulate the immune response. The encapsulation and release of pneumococcal surface protein A from four polyanhydride chemistries based on sebacic acid (SA), 1,6-bis-(*p*-carboxyphenoxy)hexane (CPH) and 1,8-bis-(*p*-carboxyphenoxy)-3,6-dioxaoctane (CPTEG) demonstrated sustained drug delivery over a month. The protein released from amphiphilic chemistries was shown to retain its structure through a combination of SDS-PAGE and circular dicroism. The antigenicity of protein released from amphiphilic polyanhydride nanoparticles was preserved as confirmed with ELISA. We also assessed the preservation of the biological functionality of PspA through a functional assay which takes advantage of the ability of secretory PspA to bind to apolactoferrin and prevent its bactericidal effects. In these studies, we used a PspA deficient strain of *streptococcus pneumoniae* and demonstrated that PspA released from amphiphilic polyanhydride nanoparticles retained the ability to bind to apolactoferrin and prevent the killing of PspA deficient *S. pneumoniae*. Together, these studies provide a framework for the rational design of a pneumonia vaccine based on amphiphilic polyanhydride nanoparticles.

(550d) Oral Mucosal Vaccination Using Coated Microneedles

Yunzhe Ma, Wenqian Tao and Harvinder Gill, Chemical Engineering, Texas Tech University, Lubbock, TX

The oral cavity has significant potential as a mucosal vaccination site because it is easily accessible and is immunologically attractive due to the presence of several lymphoid tissues in the oral cavity that could potentiate a robust immune response. Despite these advantages, the oral cavity remains under-utilized for vaccination due to the difficulty of delivering vaccine antigens across the tough transport barrier formed by the stratified epithelium covering the oral cavity tissues and the poor residence time due to salivary flow. We present the first study demonstrating the use of coated microneedles for enhancing vaccine delivery to the oral cavity using a rabbit animal model. In this study, we tested the ability of coated microneedles to penetrate the oral cavity tissues, specifically the inner lower lip and the dorsal surface of the tongue. Penetration of microneedles into tissues was confirmed by histological analyses. Delivery efficiency of coatings was found to be 63.9% and 91.2% for the lips and tongue respectively despite the moist environment in the oral cavity. Ovalbumin-coated microneedles were inserted into the lip and the tongue of rabbit to evaluate the immune response by direct delivery of antigens to the oral cavity tissues. Significantly higher ($p<0.05$) ovalbumin-specific IgG antibodies were found in serum. However, the increase in mucosal IgA antibody secretions was marginal ($p>0.05$). In addition, no significant difference ($p>0.05$) was observed between the lip or the tongue as the site for vaccination. In conclusion, this study provides the first validation of the use of microneedles for delivery to the oral cavity with specific focus on vaccination to induce both systemic and mucosal immunity using antigen-coated microneedles as minimally invasive devices for oral cavity vaccination. This methodology could form the basis of delivery of safe adjuvants to the oral cavity to further potentiate mucosal immunity.

(550e) Transdermal Delivery of Biopharmaceuticals Using Microsecond Thermal Ablation

Jeong Woo Lee, Chemical and Biomolecular Engineering, Georgia Institute of Technology, Atlanta, GA, Priya Gadiraju, Georgia Institute of Technology, Jung-Hwan Park, Bionano Technology, Gachon University, Seongnam, South Korea, Mark G. Allen, Georgia Institute of Technology, Atlanta, GA and Mark R. Prausnitz, School of Chemical and Biomolecular Engineering, Georgia Institute of Technology, Atlanta, GA

SEQ CHAPTER \h \r 1**TRANSDERMAL DELIVERY OF BIOPHARMACEUTICALS USING MICROSECOND THERMAL ABLATION**

Jeong Woo Lee[1], Priya Gadiraju[1], Jung-Hwan Park[3], Mark G. Allen[1,2], Mark R. Prausnitz[1]

[1]*School of Chemical and Biomolecular Engineering, Georgia Institute of Technology, Atlanta, GA*

[2]*School of Electrical and Computer Engineering, Georgia Institute of Technology, Atlanta, GA*

[3]*Department of BioNano Technology and Gachon BioNano Research Institute, Gachon University,*

Sungnam, Republic of Korea

Introduction

The conventional transdermal patch has provided advantages of non-invasive and simple administration, but it is generally limited to only small molecular weight and lipophilic therapeutics due to the protective function of the skin, which mostly resides in the top superficial layer of epidermis, stratum corneum. Macromolecular biopharmaceuticals cannot be delivered through the skin at therapeutic rates due to the stratum corneum barrier, so that various technologies have been investigated to overcome the skin barrier property such as mechanical disruption, radio frequency-based heating, ohmic heating, and laser ablation. The key to success with skin ablation technology is to develop methods to remove the stratum corneum barrier selectively without causing deeper skin damage [1]. Here, we present transdermal delivery of biopharmaceuticals with microsecond thermal skin ablation technology assessed by modeling of heat penetration into the skin, fabrication and characterization of a novel thermal ablation device, and skin permeability measurement.

Methods and Materials

Heat penetration into the skin was simulated using ANSYS software employing a one-dimensional transient semi-infinite model with the boundary condition [2] that the temperature of the skin surface was held constant at various temperatures up to 1100 ° during the thermal exposure of 1 μs to 100 ms. Simulations were carried out with and without a conductive mask used for guiding thermal energy only and with and without a "windows" mask used for localized ablation effect as shown in Fig. 1.

The thermal ablation device was designed by modifying a laminated gas generator actuator developed previously for aeronautical purposes [3]. A microchamber consisting of five patterned and integrated layers (two brass electrodes, two PMMA backing layers, and one microchamber layer) was fabricated using laser micromachining and lamination of each layer as shown in Fig. 2A. Two different mask designs were fabricated to facilitate mechanistic study and device optimization. A conductive mask was designed to efficiently conduct heat without allowing physical contact between the steam and the skin. A windows mask was designed to allow heat transfer to the skin through 100 μm-diameter holes in the mask and to insulate the skin everywhere else.

For the time characterization of thermal ablation, two voltage driver circuits were placed on either side of the microchamber to measure the voltage drop across the microchamber and the MOSFET switch. A ring-type piezoelectric force sensor was attached behind the microchamber to measure the force exerted by the actuation of the device. During device actuation, in some cases a high-speed video was taken at a rate of 100,000 frames per second (EKTAPRO 4540, Kodak).

To visualize the selective and localized removal of stratum corneum, a piece of full-thickness pig cadaver skin prepared with IACUC approval was ablated and then a hydrophilic sulforhodamine solution was applied to the skin surface. To measure the permeability of thermally abated skin, human cadaver skin was obtained with approval from the Georgia Tech Institutional Review Board and either left intact or exposed to thermal ablation. The skin permeability was measured for up to 48 h using a Franz diffusion cell, the donor compartment of which was filled with either 10^{-3} M sulforhodamine B or 10^{-4} M Texas Red-labeled bovine serum albumin (BSA) in PBS.

Results and Discussion

Our goal was to identify the longest thermal exposure that does not allow significant heat to reach the viable epidermis using finite element modeling. In Case I shown in Fig. 1A, when the skin is directly exposed to superheated steam at 1100 °, the viable epidermis is predicted to have no significant temperature rise after exposures of 1 ms, 10 ms and 100 ms as shown in Fig. 1B. Because heat transfer is most delayed in the stratum corneum layer with the lowest thermal conductivity of all the skin layers, heat barely penetrates into the viable epidermis during these short exposures. After a 1 ms exposure, the viable epidermis is predicted to be heated to a temperature of 180 °, which is expected to injure cells and cause pain as shown in Fig. 1E.

The mask design used in this study as in Case II and III shown in Fig. 1A permits sufficient heat transfer that the skin surface temperature is almost the same as the ejectate temperature after millisecond exposures, but is significantly lower after microsecond exposures as shown in Fig 1C and 1D. These data suggest that heating the skin for on the order of 100 ms may be optimal because that exposure time is long enough to allow the skin surface to be substantially heated, but still short enough to avoid heating below the stratum corneum.

SHAPE * MERGEFORMAT

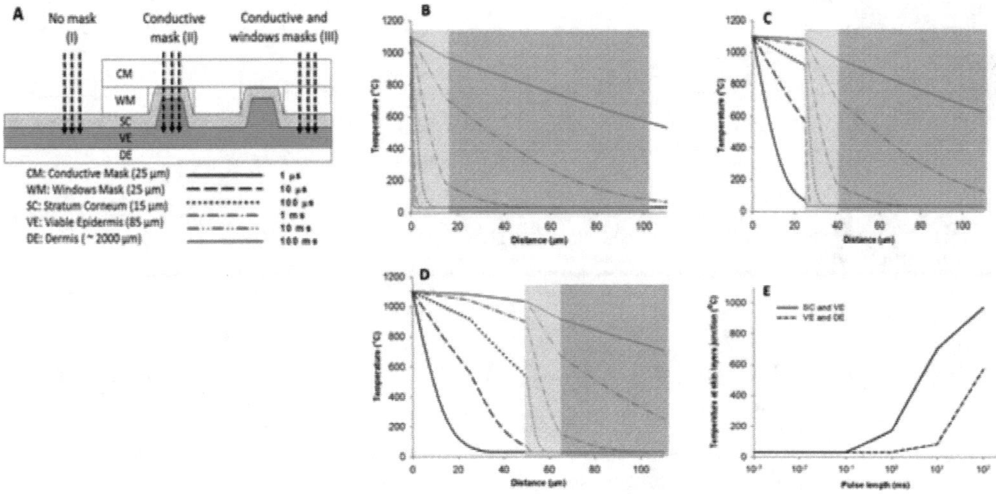

Figure 1. Simulation of heat penetration across skin heated on its surface to 1100°C for 1 ms to 100 ms. (A) Schematic showing skin heating with different schemes. Simulation predictions for heating (B) as direct exposure, (C) through a conductive mask and (D) through a conductive and a windows mask. (E) The time required to increase skin temperature by 10°C at the interface.

Guided by the heat transfer simulation, we designed a thermal ablation device to achieve thermal exposures on the order of 100 µs as shown in Fig. 2. After the microchamber is connected to a power supply circuit (Fig. 2B), the electrical energy stored in the capacitor (Fig. 2C) is discharged rapidly with the MOSFET switch on, generating an electrical discharge between the two electrodes in the microchamber. This nearly instantaneous event heats, vaporizes and ejects the saline from the microchamber through the nozzle, indicating the rapid transformation of electrical energy into thermal and mechanical energy that can be used for skin ablation.

We found that the thermal ablation device operates on a timescale on the order of 100 µs. The video image shown in Fig. 3A is the image of ejectate captured in a single frame, indicating that the steam jet was ejected within 10 µs at the most. As shown in Fig. 3B, the voltage across the microchamber rapidly rises and peaks at 7 µs, which is followed by a sudden voltage drop until 50 µs, which is the primary electrical discharge occurring in the microchamber. As shown in Fig. 3C, the force exerted during electrical discharge peaks at 100 µs, which corresponds to the timescale over which the primary steam ejection occurs. The force curve then oscillates due to resonance of the apparatus and dissipation of the reaction force energy, but is not believed to reflect forces generated directly by the steam jet ejection.

Figure 2. Microsecond thermal ablation system. (A) Schematic diagram of the disintegrated layers of the microchamber, (B) Image of a microchamber connected to a power supply circuit, (C) Diagram of the electrical circuit used to transform electrical energy into thermal and mechanical energy in the microchamber

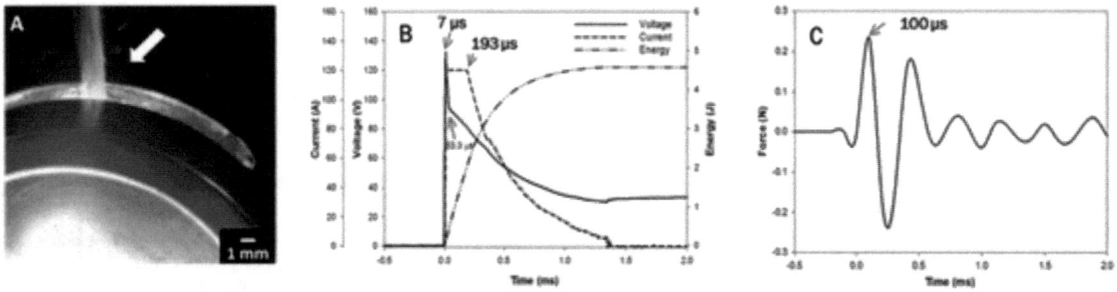

Figure 3. (A) View of the ejectate from the device, (B) Voltage change across the microchamber, and (C) Force measured during electrical discharge.

Guided by the design, fabrication, and characterization of microsecond thermal ablation system, we ablated the skin to cause selective removal of the stratum corneum as shown in Fig 4A and 4B. The selective removal of stratum corneum led to a remarkable increase in the transdermal flux of both molecules studied, as shown in Fig 4C and 4D. After a lag time, skin permeability to sulforhodamine and BSA was 3.1 (± 0.9) • 10^{-2} cm/h and 3.6 (± 0.8) • 10^{-3} cm/h, respectively. This corresponds to a 10^4-fold increase in permeability to sulforhodamine and a 10^3-fold increase in permeability to BSA after thermal ablation in comparison with intact and untreated skin.

SHAPE * MERGEFORMAT

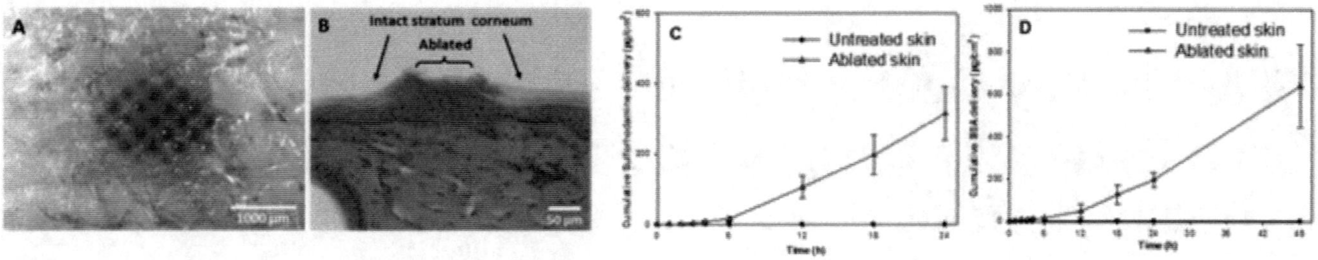

Figure 4. (A) The surface of skin ablated with microsecond thermal ablation device using a windows mask, (B) Histological image of the skin sample in (B), Cumulative delivery of (C) sulforhodamine and (D) Texas Red-labeled BSA.

Conclusion

Through modeling, we determined that thermal ablation should occur on the 100 µs time scale in order to ablate the stratum corneum selectively. To generate the high temperatures needed for skin ablation on this time scale, we designed a microdevice that rapidly heats water by an electrical discharge, thereby ejecting a superheated steam jet at the skin surface. We found that this device was able to selectively ablate stratum corneum and, when combined with a windows mask, enabled three-dimensional control over tissue removal, resulting in the increase of skin permeability to sulforhodamine and BSA. This study was funded by NIH and by Tyco Healthcare (now Covidien).

Reference

[1] J. W. Lee, P. Gadiraju, J. H. Park, M. G. Allen, M. R. Prausnitz, *J Control Release* **2011**, *154*, 58-68.

[2] M. L. Cohen, *Journal of Investigative Dermatology* **1977**, *69*, 333-338.

[3] B. English, in *School of Mechanical Engineering*, Vol. Doctor of Philosophy, Georgia Institute of Technology, Atlanta **2006**.

(550f) 28-Day Ocular Delivery of Brimonidine Tartrate From Rationally Designed Degradable Microparticles In a Rabbit Model

Morgan Fedorchak[1,2,3,4], Jeremy Wingard[2], Carlos A. Medina[2], Eiyass Albeiruti[2], Joel S. Schuman[2,3,4,5] and Steven R. Little[1,4,5,6], (1)Chemical Engineering, University of Pittsburgh, Pittsburgh, PA, (2)Ophthalmology, University of Pittsburgh, Pittsburgh, PA, (3)Louis J. Fox Center for Vision Restoration, Pittsburgh, PA, (4)McGowan Institute for Regenerative Medicine, Pittsburgh, PA, (5)Bioengineering, University of Pittsburgh, Pittsburgh, PA, (6)Immunology, University of Pittsburgh, Pittsburgh, PA

Glaucoma is the second leading cause of blindness worldwide, expected to affect up to 3 million Americans by 2020 (Quigley HA. Lancet. 2011;377[9774]:1367-77). One of the main risk factors in glaucoma is an unsafe increase in intraocular pressure (IOP). IOP reduction in patients with glaucoma is typically accomplished through the administration of medicated eye drops several times daily, the difficult and frequent nature of which contributes to patient compliance rates estimated to be as low as 50% (Hermann MM. Int Ophthalmol. 2010;30[4]:385-90). Medications such as brimonidine tartrate (BT), which requires dosing every 8-12 hours, have yet to be adapted into controlled-release formulations that could drastically improve compliance. The purpose of this study was to develop and test a controlled release BT formulation that provides one month of therapeutic levels of drug, thereby eliminating compliance issues inherent to eye drop administration.

BT was loaded into poly(lactic-co-glycolic) acid (PLGA) microparticles, which were fabricated using a standard double emulsion procedure. Both blank and drug-loaded microparticles were characterized for surface morphology and volume average diameter. In vitro release of BT from the drug-loaded microparticles was quantified by incubating a known mass of microparticles in buffer and measuring the absorption using a spectrophotometer. For our in vivo studies, groups of 3 healthy Dutch belted rabbits were randomized to receive blank microparticles (no drug), BT-loaded microparticles, or 0.2% BT drops (Alphagan¨, Allergan, Irvine, CA). Rabbits in both microparticle groups received a subconjunctival injection of microparticles suspended in sterile saline on Day 0 in one eye only. Rabbits in the positive control group received a single drop of Alphagan¨ solution in one eye twice a day each day of the study. Intraocular pressure was monitored over 28 days in both eyes and slit lamp examination was performed periodically to assess the effect, if any, of the microparticles on the surrounding tissue. Following sacrifice on Day 28, both eyes were enucleated for histological analysis. All slides were masked prior to performing histopathological analysis for evidence of irritation or foreign body response.

Microparticles were confirmed to have a diameter of 7.5±2.9 μm with a primarily poreless morphology. They released an average of 0.62±0.34 μg BT/mg particles/day in our in vitro setup, within the calculated therapeutic range of Alphagan¨ drops. Our in vivo study of the BT-loaded microparticles demonstrated that the decrease in IOP was significantly lower ($p<0.05$) in the treated eye for BT drops versus BT microparticles for all time points. There was no significant difference in ΔIOP for the untreated eye between BT drops and BT microparticles. In contrast, IOP steadily increased in rabbits injected with the blank microparticles. The microparticle bleb was visible using light microscopy and no evidence of microparticle migration was seen. Histological analysis showed no signs of foreign body response due to the microparticles or migration of the microparticles from the periocular area.

The BT-loaded microparticles presented in this study are capable of releasing a therapeutically relevant amount of a common glaucoma medication for over 30 days. Additionally, treatments based on BTMPs demonstrate a significantly greater drop in IOP over the entire 28-day study compared to the baseline IOP in a healthy rabbit model. Subconjunctival injections of the microparticles resulted in neither foreign body response nor infection after the full 28-day experimental period, and microparticles were confined to the injection area. We are currently adapting this controlled-release BT delivery system into a simple, patient-administered formulation that provides the same IOP reduction potential and biocompatibility without the need for intraocular injections.

(550g) Preclinical Evaluation of Treg Recruiting Microparticles for the Treatment of Periodontitis

Andrew J. Glowacki[1,2,3], Sayuri Yoshizawa[4,5,6], Siddharth Jhunjhunwala[7], Gustavo P. Garlet[8], Charles Sfeir[2,4,6] and Steven R. Little[1,7,9,10,11], (1)Chemical Engineering, University of Pittsburgh, Pittsburgh, PA, (2)The Center for Craniofacial Regeneration, University of Pittsburgh, Pittsburgh, PA, (3)McGowan Institute for Regenative Medicine, University of Pittsburgh, Pittsburgh, PA, (4)Oral Biology, University of Pittsburgh, (5)The Center for Craniofacial Regeneration , University of Pittsburgh, (6)McGowan Institute for Regenative Medicine, University of Pittsburgh, (7)Bioengineering, University of Pittsburgh, Pittsburgh, PA, (8)Department of Biological Sciences, School of Dentistry of Bauru, University of Sao Paulo (FOB/USP), Bauru, Brazil, (9)McGowan Institute for Regenerative Medicine, Pittsburgh, PA, (10)Immunology, University of Pittsburgh, Pittsburgh, PA, (11)The Center for Craniofacial Regeneration, University of Pittsburgh

Periodontal disease (gum disease) affects an estimated 78 million Americans, and is the leading cause of tooth loss according to the American Dental Association[1]. Current treatments for periodontal diseases focus on the removal of invasive bacterial species through non-surgical scaling and root planing procedures (sometimes accompanied by antibiotics)[2]. Although bacteria initiate the disease, tissue destruction in the periodontium (ligament and tooth supporting bone) is perpetuated by the inflammatory immune response. Recently, a subset of lymphocytes known as regulatory T cells (Tregs) has been shown to play a critical role in the regulation of harmful inflammatory immune responses[3]. Furthermore, the presence or absence of Tregs has been correlated with the presence or absence of aggressive periodontitis[4]. Therefore, we hypothesized that resolution of inflammatory conditions through the recruitment of Tregs may alleviate periodontal disease symptoms and prevent alveolar bone loss.

In support of this hypothesis, here we show Treg recruitment to poly(lactic-co-glycolic) acid microparticles encapsulating CCL22 placed in the subgingival pocket, lead to the resolution of periodontal disease symptoms in both a murine and canine model. Specifically, upon injection of CCL22 loaded microparticles in the mouse model, we observed an increase in the expression of FOXP3 (Treg transcription factor), and anti-inflammatory cytokines IL-10 and TGFβ. Indeed, we also observed a corresponding decrease in the expression of inflammatory cytokines IL-1, TNF, and IFNγ, without affecting the protective host response (we observed comparable bacterial loads, Fig. 1A). Furthermore, in the most widely accepted preclinical model for periodontitis (dog model), we observed decreased clinical scores for inflammation, probing depth (Fig. 1B), and bleeding on probing. In addition, we observed a marked decrease in alveolar bone loss in the CCL22 treated animals (observed and quantified using microCT, Fig. 1C).

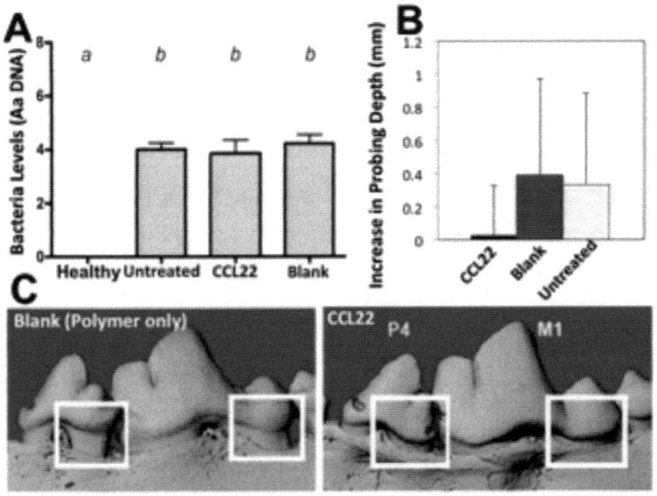

Figure 1: CCL22 Microparticles reduce periodontal disease symptoms, and maintain a protective host immune response. **(A)** Periodontal pathogen *Actinobacillus actinomycetemcomitans* DNA extracted from periodontium of mice. **(B)** Periodontal probing depths 4 weeks after disease induction in beagle dogs. **(C)** MicroCT 3-D reconstruction of the 4th premolar and carnassial of the left mandible of beagle dogs.

[1] P. I. Eke, R. J. Genco, *Journal of Periodontology* **2007**, 78, 1366.

[2] R. C. Williams, D. W. Paquette, S. Offenbacher, D. F. Adams, G. C. Armitage, K. Bray, J. Caton, D. L. Cochran, C. H. Drisko, J. P. Fiorellini, W. V. Giannobile, S. Grossi, D. M. Guerrero, G. K. Johnson, I. B. Lamster, I. Magnusson, R. J. Oringer, G. R. Persson, T. E. Van Dyke, L. F. Wolff, E. A. Santucci, B. E. Rodda, J. Lessem, *Journal of Periodontology* **2001**, 72, 1535.

[3] Q. Tang, J. A. Bluestone, *Nature Immunology* **2008**, 9, 239.

[4] G. P. Garlet, C. R. Cardoso, F. S. Mariano, M. Claudino, G. F. de Assis, A. P. Campanelli, M. J. Avila-Campos, J. S. Silva, *Journal of Clinical Periodontology* **2010**, 37, 591.

(550h) Development of Two Steps Tissue Allograft with Sequential Growth Factor Delivery to Promote Revascularization in Denudated Pneumonectomy Bronchial Stump

Eva M. Martín del Valle[1], Cristina Rodriguez Rivero[1], Gonzalo Varela[2], Dolores Ludeña[3], Marta Regueiro-Purriños[4] and Miguel A. Galán[1], (1)Department of Chemical Engineering, University of Salamanca, Salamanca, Spain, (2)Thoracic Surgery, Universitary Hospital of Salmanca, Salamanca, Spain, (3)Anatomic Pathology, Universitary Hospital of Salamanca, Salamanca, Spain, (4)Animal Surgery , University of Leon, Leon, Spain

Devascularization due to large lymph node dissection is one of the causative factors of bronchial stump fistula (BF). In this work we have developed and analyzed the utility of a tissue allograft composed of alginate polymer (AP), autologous fibrin (AF), tissue-engineered fibroblasts (TEF), vascular endotelial growth factor (VEGF) and encapsulated fibroblast growth factor (FGF) to prevent BF and promote revascularization of the bronchial stump in a pig pneumonectomy model. To validate the system a left pneumonectomy was performed in 15 pigs, leaving a long and exhaustively devascularized bronchial stump. The animals were allocated in groups of five to receive no bronchial stump coverage (group A), coverage with AP + AF only (group B) or coverage with AP + AF + TEF + VEGF + FGF (group C). After being euthanized on postoperative days 3,7, 14 , 21 and 30, bronchial stumps for each group were tested for BF occurrence and subsequently sectioned for histology and stained with hematoxylin-eosin, CD-31 and α-smooth muscle actin antibodies to evaluate blood vessels formation and maturation, capillary density and TEF proliferation by means of a semiquantitative method.

From the analysis done it can be concluded that: No BF could be detected for any group, probably due to a high spontaneous healing capacity of pig tissues. In group A, fibroblastic proliferation and revascularization were scarce; group B showed some immature neovessels besides a moderate TEF proliferation while in group C, in addition to an intense TEF proliferation, a large number of vascular yolks could be detected on postoperative day 7, continuing to complete vascular structures on postoperative day 14 within a highly dense capillary network. Therefore, the avascular scaffold highly promotes revascularization of the denudated bronchial stump. Further studies are being done to clarify the clinical application of these findings.

(557a) Cancer Cell Migration in 3D

Stephanie I. Fraley[1], Yunfeng Feng[2], Pei-Hsun Wu[3], Gregory Longmore[4] and Denis Wirtz[1], (1)Chemical and Biomolecular Engineering, Johns Hopkins University, Baltimore, MD, (2)Cell Biology, BRIGHT Institute, Washington University, St. Louis, MO, (3)Chemical and biomolecular Engineering, Johns Hopkins University, Baltimore, MD, (4)Medicine, Cell Biology, BRIGHT Institute, Washington University, St. Louis, MO

Although much is known about mechanical and spatial signal transduction in cells migrating on conventional two-dimensional (2D) substrates, drastic functional differences are apparent in the molecular machinery driving cell migration through more physiologically-relevant, fibrilar, three-dimensional (3D) environments. Taking an inside-out approach to understanding the function of cell adhesion proteins in 3D cell migration, we use shRNA constructs to silence expression of a variety of focal adhesion proteins and demonstrate that external dimensional cues are critical to protein localization and signaling. We show that 3D cell motility parameters are not predicted by the same parameters in 2D. Likewise, the focal adhesion protein zyxin plays a distinctive role in the temporal and dimensional aspects of cell trajectories in 3D and 1D microenvironments, but not on 2D substrates. Lastly, taking an outside-in approach, we demonstrate that changes in 3D matrix microstructure alone modulates modes of cell motility. This occurs via microstructure induced spatial positioning of protrusions and matrix metalloproteinase (MMP) activity. Thus, cell motility may only be abrogated by small molecule inhibition of MMPs in distinct matrix microstructures.

References

Fraley *et al*. Nature Cell Biology 12: 598-604 (2010)

Fraley *et al*. Nature Cell Biology 13: 5-7 (2011)

Fraley *et al*. Nature Communications 3:719-731 (2012)

(557b) Engineering Chimeric Antigen Receptors Targeting an Endogenous Murine Tumor Associated Antigen

Cary F. Opel[1], Matthias T. Stephan[2] and K. Dane Wittrup[1,3], (1)Chemical Engineering, Massachusetts Institute of Technology, Cambridge, MA, (2)Clinical Research Division, Fred Hutchinson Cancer Research Center, Seattle, WA, (3)Biological Engineering, Massachusetts Institute of Technology, Cambridge, MA

Adoptive cell transfer (ACT) therapy for the treatment of cancer is the act of administering lymphocytes to a patient with the goal of achieving an active immune response against tumor cells. A variety of methods have emerged to modify the lymphocytes in order to enhance their anti-tumor potency. One increasingly effective method is the use of chimeric antigen receptors (CAR) to target the lymphocytes specifically to tumor associated antigens. CARs are surface bound proteins consisting of an intracellular activation domain fused to an extracellular targeting domain. These receptors enable the lymphocytes to bypass many of the restrictions on the endogenous activation pathway, potentially allowing for a more effective immune response to the target tumor cells.

Despite many recent successes using lymphocytes expressing CARs in a clinical setting, some unforeseen side effects have been reported, and the design criteria for the most effective CAR domain structure have yet to be clearly established. Of critical importance to the design of a CAR is the selection of internal signaling domains. Weakly activating domains may not elicit a strong enough response to eradicate the cancer, while potent activating domains may cause an autoimmune response with severe side effects. Additionally, some activating domains may enhance other beneficial properties such as the survival, persistence, or proliferation of the transferred lymphocytes. First generations CARs used only the activating domain of CD3, while more recent second and third generation CARs combine two or three internal activating domains to enhance the effectiveness of the transferred lymphocytes. Unfortunately, the exact benefits and side effects of the various domain combinations remain unclear. This work aims to elucidate some of the design parameters that enable the most effective targeting of tumor cells by CAR expressing lymphocytes.

To study the effects of different activating domains, a model system was developed using the poorly immunogenic murine melanoma cell line, B16F10, and an antibody, TA99, targeting gp75, an endogenous tumor associated antigen expressed on the surface of B16F10 cells. The targeting domain of a CAR was created using an scFv constructed from the variable regions of the TA99 antibody. Preliminary cytotoxic assays using a first generation CAR showed promising results, so several CARs were constructed using various combinations of internal signaling domains and their effectiveness was tested *in vitro*.

Unlike many other model systems for adoptive cell transfer (ACT), this model will be expanded to use immune competent C57BL/6 mice. Typically, the effectiveness of CARs is tested on immunodeficient mice using human cell lines and targeting human antigens. In these models, off-target effects as well as interactions with the immune system are not evaluated, creating potentially misleading results. By using a fully endogenous system, a more accurate measurement of the effectiveness of various activating domains may be achieved, providing a more accurate prediction of how treatments will perform in a clinical setting.

(557c) Genetically Modified Bacteria Actively Secrete Recombinant Protein Biomarker for Early Detection of Solid Tumors

Jan Panteli, Chemical Engineering, University of Massachussetts Amherst, Amherst, MA, Britanny Forkus, Chemical Engineering, University of Massachussetts Amherst, Amhest, MA and Neil S. Forbes, Chemical Engineering, University of Massachusetts, Amherst, MA; Chemical Engineering, University of Massachusetts, Amherst, Amherst, MA

An attenuated strain of *Salmonella typhimurium* has been engineered to actively secrete *Zoanthus sp.* green fluorescent protein (ZsGreen) in a controlled manner for use in *in vitro* and *in vivo* tumor models for early detection of cancer. Current methods of cancer detection involve a primary screen for abnormal levels of native biomarkers, which might indicate cancer and other diseases. These preliminary screens often yield false results, requiring an invasive biopsy that is often painful and discomforting, to confirm diagnosis. Secretion of an exogenous recombinant protein from within tumor tissue allows cancer detection through noninvasive blood sampling without uncertainty in the results. It has been shown previously that systemically administered *Salmonella typhimurium* localize in solid tumor tissue 2000 fold more than in healthy tissue (Low *et al.* 1999). Once colonization of the bacteria in the tumor is achieved, triggered production and secretion of a protein biomarker is transported back to the blood stream for detection. Fluid transport in tumor interstitial space has been shown to have outward flowing pressure gradients hindering systemic administration of small molecule chemotherapeutics. These transport limitations act in favor of export of protein from within tumor tissue to the blood stream. A secreted biomarker, produced from the bacteria, within tumor tissue, would indicate the presence of cancer without invasive procedures and require a simple blood draw for detection. In this study a gram negative type III secretion system(T3SS) signal was fused to the n-terminus of ZsGreen protein (MvZsGreen) yielding a 3-fold increase in secretion in wild type strain (SL1344) and 5-Fold increase in an attenuated strain (VNP20009) of *S. typhimurium*. Over half of the protein produced is secreted from the attenuated strain (pBAD_MvZsG_VNP) under control of the L-arabinose inducible promoter (pBAD). Time-lapse fluorescent microscopy with viability staining revealed that a majority of cells (85%) are viable after 36hrs of culture and release protein at twice the rate than dead cells. Immunoblotting measurements and fluorescence measurements of secreted recombinant MvZsGreen diluted in blood was shown to have detection limits of 1:10 and 1:1000 respectively. A previously developed micro-fluidic 4T1 tumor model (Toley *et al.* 2011) was used to determine transport rates of protein through tissue and detection limit on the secreted protein from actual tumor tissue under triggered induction using the L-arabinose inducible system. Increased secretion from VNP20009 over the wild type *S.* typhimurium is extremely beneficial in reducing virulence while improving protein production and transport for detection. Viability studies revealed two mechanisms of protein release in culture with an increased rate in active secretion through T3SS compared to cell death related diffusion. This allows production and secretion of biomarker, MvZsGreen, as the bacteria proliferate within the tumor tissue, which is key for monitoring cancer growth for prolonged periods of time. MvZsGreen has higher sensitivity for detection via fluorescence than immunoblotting allowing for more accurate determination of the presence of a tumor mass. Micro-fluidic studies revealed favorable MvZsGreen transport through solid tumor tissue and subsequent detection. This system demonstrates the ability of bacteria to produce recombinant protein in a controlled manner deep within tissue, secrete that protein, and have the protein transport through tumor tissue to the blood stream for detection. These studies demonstrate the feasibility of a engineered bacterial tool for early detection and diagnosis of cancer.

(557d) A Decision Tool for the Design of Optimal Personalised Chemotherapy Protocols for the Treatment of Acute Myeloid Leukaemia (AML)

Eleni Pefani, Chemical Engineering, Imperial College of London, London, United Kingdom, Nicki Panoskaltsis, Department of Haematology, Northwick Park Hospital, London, United Kingdom, Athanasios Mantalaris, Centre for Process Systems Engineering, Department of Chemical Engineering, Imperial College, London, United Kingdom, Michael C. Georgiadis, Centre for Process Systems Engineering, Chemical Engineering, Imperial College London, London, United Kingdom and Efstratios N. Pistikopoulos, Chemical Engineering, Imperial College London, Centre for Process Systems Engineering, London, United Kingdom

AML is an aggressive cancer of the blood and bone marrow resulting in failure of the blood and immune systems. Immature white blood cells which are not able to develop into normal functioning blood cells are overproduced and build up in the bone marrow and blood. This inhibits the development of healthy blood and immune cells due to space restrictions and inhibitory and clonal factors specific to the disease (Panoskaltsis, 2003; Panoskaltsis, 2005).

The most common treatment for AML is intensive chemotherapy requiring high doses of anticancer drugs to eradicate cancer cells. Although effective for debulking of the leukaemia tumor burden, the use of these chemotherapy regimens is also highly destructive to normal tissues and cells, often to a life-threatening extent. Currently, the choice of chemotherapy drugs, dose and schedule only take few patient-specific factors into consideration and the choice of treatment often depends on the treating physician's experience. However, a more systematic approach for the design of treatment protocols that will depend on specific inter-patient and intra-leukaemia factors is required. This systematic treatment design will lead to drug individualization and treatment dose optimisation so as to balance the benefits of higher dose therapy against the increase in toxicity in normal tissue. As presented in previous work (Dua et. al., 2005; Dua et. al., 2008; Pefani et. al. 2011), this can be achieved by modelling the key biological mechanisms as a means to gain insight into the chemotherapy procedure, which can then be used as a predictive tool for patient response during treatment.

In this work, a mathematical model is derived by combining the actions on the cell cycle, (drug target), with pharmacokinetic and pharmacodynamic aspects. This combination should provide a comprehensive description of drug diffusion and action after administration. The derived model depends on patient characteristics such as age, sex, body mass index and on disease characteristics such as tumor burden at the point of diagnosis and individual AML cell cycle characteristics.

Simulation and optimisation results are presented using the derived model for two patient case studies. The studied patients are of same sex, age and body mass index characteristics. However, their initial tumour burden and cell cycle times differ. They are both treated with the same treatment protocol in the model as they would be in clinical practise (a combination of cytarabine and daunorubicin). At completion of the first chemotherapy cycle one patient displays a bone marrow which contains a higher number of normal cells, whilst the other exhibits a higher number of AML cells. An optimisation problem is formulated thereafter for the latter patient resulting in the normal cell population incrementing enough to allow for normal bone marrow tissue recovery prior to the next chemotherapy cycle to further decrease AML cells.

Model analysis results of these two patient case studies reveal the potential future applicability of the proposed tool and indicate the need for an interdisciplinary approach between clinicians, experimental- and model-based experts to collaborate in enabling personalised schedules for chemotherapy delivery in cancers such as AML.

References

Panoskaltsis N, Reid CDL, Knight SC. Quantification and cytokine production of circulating lymphoid and myeloid cells in acute myelogenous leukemia (AML). *Leukemia* 2003; 17:716-730.

Panoskaltsis N. Dendritic cells in MDS and AML – cause, effect or solution to the immune pathogenesis of disease? *Leukemia* 2005; 19:354-357.

Dua P., Dua V., Pistikopoulos E. N.: *Optimal delivery of chemotherapeutic agents in cancer*. Computers & Chemical Engineering 32(1-2): 99-107 (2008).

Dua P., Pistikopoulos E. N.: *Modelling and control of drug delivery systems*. Computers & Chemical Engineering 29(11-12): 2290-2296 (2005).

Pefani E., Panoskaltsis N., Mantalaris A., Georgiadis M. C., Pistikopoulos E. N.. Modelling and Simulation of Drug Delivery Systems for the treatment of Acute Myeloid Leukemia. Proceedings of the 5th European Conference of International Federation for Medical and Biological Engineering (IFMBE), Budapest, Hungary, 2011.

Acknowledgment

This work is supported by European Research Council (MOBILE, ERC Advanced Grant, No: 226462), The Richard Thomas Leukaemia Fund and the CPSE Industrial Consortium.

(557e) Pharmacokinetic/Pharmacodynamic Model Predicts the Response to Cancer Therapeutics Targeting VEGF

Stacey D. Finley and Aleksander Popel, Biomedical Engineering, Johns Hopkins University, School of Medicine, Baltimore, MD

Angiogenesis, the formation of new blood vessels from pre-existing vasculature, is a complex biological process involved in physiological conditions, as well as in pathological cases such as cancer. Vascular endothelial growth factor (VEGF) is a key regulator of angiogenesis and has been targeted in order to inhibit angiogenesis and prevent tumor vascularization. This study applies a molecular-detailed compartment model of VEGF kinetics and transport in the human body [1, 2] to investigate the effects of anti-angiogenic therapies targeting the VEGF pathway.

The model includes normal tissue (represented by skeletal muscle), blood, and tumor. We include molecular interactions between two major VEGF isoforms $VEGF_{121}$ and $VEGF_{165}$, receptors VEGFR1 and VEGFR2, and co-receptors NRP1 and NRP2. Macromolecular transport between compartments occurs via transendothelial microvascular permeability and lymphatic flow. Additionally, free VEGF is subjected to protein degradation in the tissue compartments, and species are cleared from the blood. VEGF receptors and co-receptors are localized on the abluminal and luminal endothelial surfaces, as well as on parenchymal cells. Receptor density is based on *in vitro* and *in vivo* quantitative flow cytometry [1, 3]. The model is represented by 125 ordinary differential equations and has been validated against experimental data.

We apply the model to predict the effects of therapeutics that inhibit VEGF or its receptors. We also investigate combination therapies and targeted delivery of the anti-angiogenic agents. The response to anti-angiogenic treatment is characterized by the level of free VEGF in the body and the number of VEGF/VEGFR2 complexes (considered to be pro-angiogenic). Following neutralization of VEGF with an antibody, plasma free VEGF and the number of VEGF/VEGFR2 complexes in the blood increase tenfold. The increase in plasma VEGF has been observed clinically. Interestingly, the model predicts that VEGF in the tumor interstitium can increase or decrease following administration of the VEGF antibody, depending on properties of the tumor microenvironment, including the relative rate at which $VEGF_{121}$ and $VEGF_{165}$ are secreted by the tumor and the expression level of the VEGF receptors on tumor cells. Additionally, we find that targeting VEGFR2 is effective in inhibiting the formation of pro-angiogenic complexes, while mediating the concomitant increase in free VEGF that can occur with inhibition of VEGF or VEGFR1.

We have developed a framework for investigating the systemic effects of anti-angiogenic drugs that inhibit VEGF-mediated signaling, which depend on tumor-specific properties and are difficult to quantify clinically. The model should be instrumental in optimizing personalized cancer treatment strategies that target the VEGF pathway.

This work was supported by NIH grant R01 CA138264 (ASP), NIH fellowship F32 CA154213 (SDF), and UNCF/Merck Postdoctoral Fellowship (SDF).

References

1. Finley, S.D., et al., *Pharmacokinetics and pharmacodynamics of VEGF-neutralizing antibodies.* BMC Systems Biology, 2011. **5**:193.

2. Finley, S.D. and A.S. Popel, *Predicting the effects of anti-angiogenic agents targeting specific VEGF isoforms.* AAPS Journal, 2012. In press.

3. Imoukhuede, P.I. and A.S. Popel, *Quantification and cell-to-cell variation of vascular endothelial growth factor receptors.* Experimental Cell Research, 2011. **317**(7):955-965.

An integrated systems-based modelling framework for investigating the effect of anticancer drugs on solid tumours

Cong Liu, J. Krishnan, Xiao Yun Xu

Department of Chemical Engineering, Imperial College London, UK

Abstract

Background. The effectiveness of clinical chemotherapy is dependent on the penetration of anticancer agents in tumour tissues and the responsiveness of tumour cells towards exposure to administered drugs. Most blood-borne anticancer drugs gain their access to tumour cells by going through a series of physical and biological processes, including convection in the tumour vasculature, extravasation across the capillary wall, penetration through the tumour interstitium and finally translocation inside tumour cells. However, these transport processes are often impaired by the abnormal tumour microenvironment (i.e. irregular tumour vasculature, elevated interstitial fluid pressure, and extracellular matrix composition), resulting in limited drug concentration available at the target site. Upon interaction with their intracellular targets, anticancer drugs induce multiple cellular signalling pathways, through which they exert therapeutic effects with apoptosis (programmed cell death) being of primary interest. The way by which cellular signalling functions is extremely complex as cellular pathways are carried out in a complex and interconnected network and usually display highly non-linear input-output relationships. Given the complexities associated with drug transport in tumour tissues and intracellular signal transduction, understanding the effect of anticancer drugs on solid tumours presents a highly challenging problem.

Methods. In this study, an integrated systems-based modelling framework is developed which is capable of (i) capturing relevant biological processes and their interconnections, and (ii) providing predictive and mechanistic insight into the effects of chemotherapeutic agents. As a first attempt, the model starts with basic descriptions of the essential elements with a focus on their integration in the system. The essential modules include drug transport, intracellular apoptosis signalling and tumour cell density dynamics.

Drug transport in tumour tissue is described by a diffusion-convection-reaction equation, with doxorubicin as the anti-cancer drug for modelling purpose. Velocity field is obtained by solving tumour blood flow, which is governed by the Navier-Stokes equations in the tumour vasculature, Starling's law for flow across the capillary wall and Darcy's law for flow in the tumour interstitium. Intracellular drug concentration serves as the stress signal triggering apoptosis pathway. Here, two types of coarse-grained intracellular apoptosis models, monostable and bistable apoptosis models, extracted from the systems biology literature are examined. Both simplified models capture key features of apoptosis- threshold effect and irreversibility. The cellular response is reflected on the tissue level in terms of tumour cell density which is described by a sharp decrease in the tumour growth rate in the logistic formulism of tumour growth. Re-distribution of tumour cell density is examined upon different pulse drug injections. The integrated systems-based model is applied to a simplified tumour cord geometry.

Results. Simulation results show that insufficient drug transport in tumour interstitium is a major limiting factor in inducing drug effect. Despite qualitatively different dynamics of signal transduction, the monostable and bistable apoptosis models exhibit similar trends of tumour cell density distribution for most drug stimuli encountered except for a case where drug-induced apoptosis is involved in enhancement in interstitial drug penetration due to lower tumour cell density with the monostable apoptosis model. The sensitivity analysis of drug-specific properties indicates that (a) there exists an optimal drug diffusivity to balance interstitial drug transport and the specific acquirement of apoptosis switch, and (b) higher vascular permeability does not lead to significant improvement in drug effect, demonstrating the interstitial drug transport being a limiting factor.

Conclusion. Despite a number of simplified assumptions employed, the present modelling framework is capable of providing solid conclusions of reasoning-consequence for some aspects of the complex problem and offers clear-cut insights into the effect of various drug injections and other contributing factors on drug efficacy. It serves as a credible platform, which is modular and transparent to allow for systematic incorporation of additional layers of complexities and is expected to bridge the gap between the progress in systems biology of cellular signalling processes and its application in improving the efficacy of anticancer drugs.

Key words: drug transport, solid tumour, intracellular signalling, systems-based modelling.

(557g) Development and Characterization of Novel, Micelle-Based Parthenolide Delivery Systems

Michael Baranello[1], Hannah Watkins[2], Craig Jordan[3] and Danielle Benoit[2,4,5], (1)Chemical Engineering, University of Rochester, Rochester, NY, (2)Department of Biomedical Engineering, University of Rochester, Rochester, NY, (3)Department of Medicine, Hematology/Oncology, University of Rochester Medical Center, Rochester, NY, (4)Department of Chemical Engineering, University of Rochester, Rochester, NY, (5)Center for Musculoskeletal Researcch, University of Rochester Medical Center, Rochester, NY

INTRODUCTION:

Parthenolide (PTL) is effective at eradicating a number of cancer cells, including leukemia stem cells (LSCs) but its low solubility in blood limits its use therapeutically[1,2]. One way to circumvent this problem is through the use of a polymeric delivery system[3]. We are utilizing core-shell micelles to enable robust PTL loading within the micelles' hydrophobic interior, and we aim to increase PTL's solubility and circulation time *in vivo* with a protective hydrophilic corona. Using reversible addition-fragmentation chain transfer (RAFT) polymerization, amphiphilic diblock copolymers of poly(styrene-*alt*-maleic anhydride)-b-poly(styrene) (PSMA-b-PS) and poly(styrene-*alt*-maleic anhydride)-b-poly(butyl acrylate) (PSMA-b-PBA) were synthesized, and subsequently assembled into core-shell micelles for PTL delivery. A variety of micelles made from PSMA-b-PS and PSMA-b-PBA diblocks of different overall molecular weights and relative hydrophobic:hydrophilic chain lengths have been characterized with respect to size, morphology, PTL loading capabilities, stability in aqueous solution, and *in vitro* cytotoxicity.

MATERIALS AND METHODS:

Diblock copolymers of PSMA-b-PS were synthesized in a one-step polymerization by providing an excess of styrene monomer. Various molar ratios (18:1, 9:1, 4:1) of styrene:maleic anhydride monomers were combined in dioxane with 4-cyano-4-(dodecylsulfanyltrithiocarbonyl) sulfanylpentanoic acid (DCT) chain transfer agent and azobisisobutyronitrile (AIBN) thermal initiator. The same RAFT chain transfer agent and thermal initiators were used to synthesize diblocks of PSMA-b-PBA in two-step polymerizations. ^{1}H NMR was used to confirm conversion of monomer and formation of polymer, and gel permeation chromatography (GPC) was used to measure polymer molecular weight and polydispersity index (PDI).

Diblock copolymers were self-assembled into micelle nanoparticles by adding water drop-wise to well-mixed polymer solutions in dimethylformamide. After self-assembly, micelle size and morphology was analyzed by dynamic light scattering (DLS) and transmission electron microscopy (TEM). Micelles were then loaded with PTL using solvent replacement techniques, and the PTL loading efficiency and capacity was determined using high performance liquid chromatography.

MV4-11 leukemia cells were cultured in untreated 24-well styrene plates at $5*10^5$ cells/mL with Iscove's Modified Dulbecco's Media (IMDM) containing 10% heat-inactivated fetal bovine serum, and 1% penicillin/streptomycin. Cells were dosed with free PTL at 2.5, 5, and 10 µM, and cell counts were taken 24 hours after the dose. The cytotoxicity profile of free PTL was compared untreated control groups, as well as PTL-loaded micelles at the same PTL concentrations.

RESULTS AND DISCUSSION:

PSMA-b-PS and PSMA-b-PBA polymers with molecular weights ranging from 18 to 48 kDa and estimated hydrophobic:hydrophilic molecular weight ratios between 0.5 – 1.2 were synthesized. All polymers self-assembled into spherical micelles with diameters between 20 - 200 nm. Interestingly, distinct morphological differences can be observed between micelles formed from diblocks of different hydrophobic:hydrophilic ratios (shown in Figure 1).

PSMA-b-PS micelles successfully loaded PTL with efficiencies ranging between 20% and 40%. These PTL-loaded micelles produced a 50% reduction in MV4-11 leukemia cell viability at PTL doses of 5 μM. Importantly, micelles in the absence of PTL exhibited no significant cytotoxicity towards these cells, when delivered at the same concentration as PTL-loaded micelles.

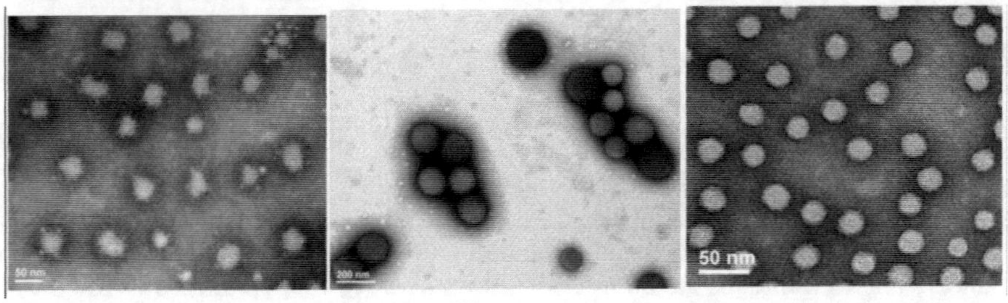

Figure 1: Representative TEM Images of Polymer Micelles. From left to right: $PSMA_{186}$-b-PBA_{107} (brush-like structures, 50 nm in diameter), $PSMA_{59}$-b-PS_{63} (spherical micelles, 200 nm in diameter), $PSMA_{100}$-b-PS_{250} (dense spherical micelles, 20 nm in diameter).

CONCLUSIONS AND FUTURE DIRECTIONS:

PSMA-b-PS and PSMA-b-PBA diblock copolymers were made using RAFT polymerization, and subsequently self-assembled into core-shell polymer micelles for drug delivery applications. A potent leukemia therapeutic, PTL, was successfully loaded into these micelles and exhibited robust LC cytotoxicity *in vitro* when delivered via polymer micelles, where unloaded micelles were non-toxic. Ongoing experiments are examining PTL release rate *in vitro*, as well as *in vivo* circulation time and biodistribution of different micelle architectures.

REFERENCES:

1. Guzman, M.L. et al. The sesquiterpene lactone parthenolide induces apoptosis of human acute myelogenous leukemia stem and progenitor cells. *Blood* 105, 4163-4169, 2005.

2. Jordan, C.T. The leukemic stem cell. *Best Pract Res Clin Hematology* 20, 13-18, 2007.

3. Torchilin, V.P. Structure and design of polymeric surfactant-based drug delivery systems. *J Control Release* 73, 137-172, 2001.

(558a) Towards Ex Vivo Platelet Manufacturing: The Importance of Matrix Elasticity and Shear Force On Megakaryocytic Differentiation

Jinlin Jiang, Stephan C. Lindsey and Eleftherios T. Papoutsakis, Department of Chemical and Biomolecular Engineering & Delaware Biotechnology Institute, University of Delaware, Newark, DE

Matrix elasticity (Engler et al., Cell **126**: 677 (2006)) and biomechanical (shear) force (Baruch et al., Blood **114**:1975 (2009)) have been identified as important differentiation cues for stem cells. In this study, we investigated how these biomechanical/biophysical signals affect megakaryocytic (Mk) differentiation from hematopoietic stem and progenitor cells (HSPCs), aiming to uncover signaling mechanisms behind these effects, and establish better culture environment for Mk differentiation and platelet biogenesis. Application of shear stress (2.5 dyne/cm^2; physiological range: 1.3~4.1 dyne/cm^2) to CHRF cells, a well-established Mk cell line, adherent to fibronectin-coated surface for two hours resulted in retraction, shedding and formation of cytoplasmic extensions (proplatelet-like structures), whereas there were no morphological changes in static culture. Primary human Mks derived from HSPCs *in vitro* also deformed their cell shape and formed proplatelet structures in response to flow application (1 dyne/cm^2). In addition, shear force accelerated endomitosis as assessed by flow-cytometry based analysis of BrdU incorporation: CHRF cells exposed for two hours to shear flow had 16% more BrdU positive cells compared to cells under static conditions. The stress responsive p53 is a regulator in both cell-cycle arrest by laminar flow (Zeng et al., JBC **278**: 24594 (2003)) and Mk differentiation (Fuhrken et al., JBC **283**: 15589 (2008)). To determine if p53 is involved in the response of CHRF cells to shear stress, Western analysis was used to show that two hours of exposure to shear (2.5 dyne/cm^2) increased the ratio of K382-acetylated p53 to total p53 by more than 188% (n=3, $p<0.05$) while total p53 protein expression did not change significantly. These data suggest that p53 acetylation is mediating, partially at least, the response of Mk cells to biomechanical forces. In order to examine the effect of matrix elasticity on Mk differentiation, polyacrylamide gels were used to control elasticity. We coated the gel surface with fibronectin to provide holding points for CHRF cells to sense gel elasticity. As matrix elasticity increased, the % of CHRF cells with cytoplasmic extensions increased first and decreased later (Fig. 1, n=3, Error bar: SEM). The highest percentage (17%) of CHRF cells displaying cytoplasmic extensions was observed on a gel with elasticity (ca. 8.2 kPa) mimicking the *in vivo* vascular niche. Non-muscle myosin IIA (NMMIIA) has been shown to serve as elasticity sensor in mesenchymal stem cells. Thus, we aimed to investigate the role of NMMIIA in the response of CHRF cells to matrix elasticity. Blebbistatin, an inhibitor of NMMIIA, minimized the differences of the percentage of cells with cytoplasmic extensions between gels with different elasticity (Fig. 2), thus suggesting that NMMIIA is involved in the response of CHRF cells to matrix elasticity and may also serve as a biological elasticity sensor. As shown by fluorescent microscopy, the cellular distribution of NMMIIA was directed by matrix elasticity (Fig. 3). On soft gels (75 Pa; n=3), NMMIIA was concentrated in the center of the cytoplasm. On a moderately stiff gel (5.6 kPa; n=3), NMMIIA was distributed across the whole cell, while on glass (the stiffest condition tested; n=3), NMMIIA moved into the nucleus and the tips of cytoplasmic extensions. Others have shown NMMIIA is involved in regulation of gene expression when it moves into the nucleus, so our observation is that NMMIIA translocates into the nucleus as gel elasticity increases, thus strongly suggesting that NMMIIA may regulate gene transcription during proplatelet formation and platelet biogenesis. We are working to further investigate and prove this hypothesis. Our data shed light on the largely unexplored role of biomechanical and biophysical parameters (of physiological significance) on Mk differentiation. The impact of these parameters is both of physiological and practical significance in the context of platelet biogenesis and ex vivo platelet generation.

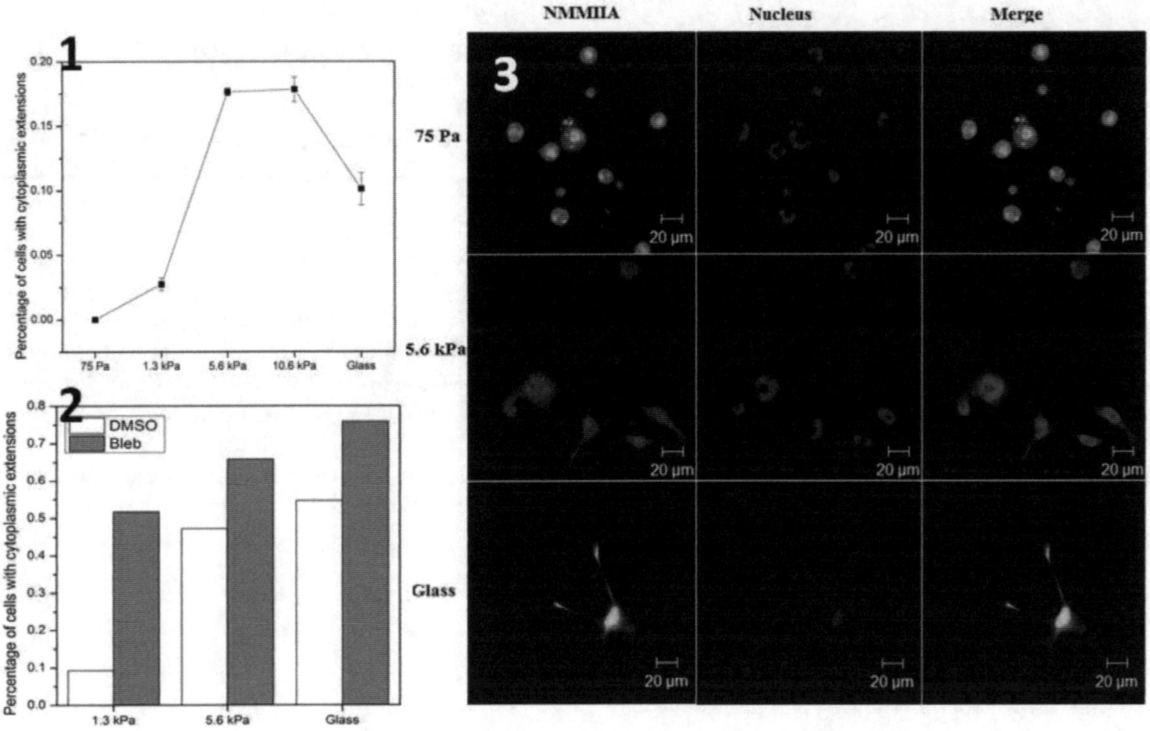
1
Percentage of cells with cytoplasmic extensions
0.20
0.15
0.10
0.05
0.00
75 Pa
1.3 kPa
5.6 kPa
10.6 kPa
Glass
2
Percentage of cells with cytoplasmic extensions
0.8
0.7
0.6
0.5
0.4
0.3
0.2
0.1
0.0
DMSO
Bleb
1.3 kPa
5.6 kPa
Glass
3
NMMIIA
Nucleus
Merge
75 Pa
5.6 kPa
Glass
20 µm

(558b) Cardiac Tissue Engineering Using Human Pluripotent Stem Cells

Tung Ying Lu, Bo Lin and **Lei Yang**, Developmental Biology, University of Pittsburgh, Pittsburgh, PA

Approximately one person dies in every 34 seconds as a consequence of heart diseases in the United States, and more than 5 million patients in the US suffer from heart failure, a disabling and frequently fatal disease condition. Without organ transplantation, end stage heart failure is irreversible. In addition, approximately 50% of heart disease patients do not respond to regular therapies including medication, device implantation and eventually require heart transplantation. Thus heart disease therapy urgently requires the development of novel disease-specific drugs, of new cell resource for replacement therapy, as well as the donor hearts for transplantation. However, those treatments largely remain underdeveloped or inadequate. Heart tissue engineering holds great promise for heart disease therapy by building patient and disease-specific engineered heart tissues for drug screening, drug testing and myocardium implantation, as well as the ultimate engineered hearts for transplantation. Recent progress in human induced pluripotent stem (iPS) cells provides the required cell resource to engineer patient-specific heart constructs for the future heart disease therapies. In this study, we engineered 3D heart tissues with human pluripotent stem cells, including human iPS cell and human embryonic stem (ES) cell derived cardiovascular cells. The engineered heart constructs exhibited spontaneous contractions and were functionally synchronized with electrical stimulation. To explore the potential pharmaceutical applications, we challenged the engineered heart constructs with various chemicals and found the constructs showed similar responses as those from the *in vivo* hearts.

(558c) Stem Cell Senescence: Nanog Reverses the Effects of Organismal Aging On Proliferation and Myogenic Differentiation Potential of Mesenchymal Stem Cells

Juhee Han[1], Panagiotis Mistriotis[2] and **Stelios T. Andreadis**[1], (1)Department of Chemical and Biological Engineering, University at Buffalo, the State University of New York, Amherst, NY, (2)Chemical and Biological Engineering, University at Buffalo, State University of New York, Amherst, NY

Although the therapeutic potential of mesenchymal stem cells (MSC) is widely accepted, loss of cell function due to donor aging or culture senescence are major limiting factors hampering their clinical application. Our laboratory recently showed that MSC originating from older donors suffer from limited proliferative capacity and significantly reduced myogenic differentiation potential. This is a major concern, as the patients most likely to suffer from cardiovascular disease are elderly. Here we tested the hypothesis that a single pluripotency associated transcription factor, namely Nanog, may reverse the proliferation and differentiation potential of BM-MSC from adult donors. Our microarray analysis showed that adult (a)BM-MSC expressing Nanog clustered close to Nanog-expressing neonatal cells. Nanog markedly upregulated genes involved in cell cycle, DNA replication and DNA damage repair and enhanced the proliferation rate and clonogenic capacity of aBM-MSC. Notably, Nanog enhanced the myogenic differentiation potential and reversed the effects of organismal aging on the contractile function of aBM-MSC to a similar level as that of neonatal (n)BM-MSC. The effect of Nanog on contractility was, in part, mediated through activation of the TGF-b pathway by diffusible factors secreted in the conditioned medium of Nanog-expressing BM-MSC. In addition, Nanog reversed senescence-dependent decline of alpha-smooth muscle actin (α-SMA) by synergizing with the TGF-b1 pathway and upregulating serum response factor (SRF). Overall, our results suggest that Nanog may be used to overcome the effects of organismal aging on aBM-MSC, thereby increasing the potential of MSC from aged donors for cellular therapy and tissue regeneration.

(558d) Multiplexed Single Cell Analysis of Embryonic Stem Cells and Induced Pluripotent Stem Cells

Jun Wang[1], Ritchie Hor[2], Kathrin Plath[2] and James Heath[3], (1)Chemistry, California Insitute of Technology, Pasadena, CA, (2)UCLA, Los Angeles, CA, (3)Chemistry and Chemical Engineering, California Institute of Technology, Pasadena, CA

Embryonic stem cells (ESCs) are derived from the inner cell mass of a developing blastocyst, propagated in *in vitro* culture systems, and defined by the ability to self-renew their pluripotent state. The ability to grow ESCs in culture relies heavily on the media components in which they grow, where specific cytokines and growth factors are necessary to maintain the self-renewal of ESCs and prevent spontaneous, terminal differentiation into heterogeneous somatic cells. Typically, human ESCs need to be grown on a layer of fibroblasts, known as feeder cells, where exogenously supplied FGF2 induces the feeders to secrete the signaling proteins TGFB1 and IGF2 to ESCs, instructing them to stably maintain their pluripotent state. ESCs can also be grown in feeder-free cultures, on the condition that the growth media are still supplemented with FGF2, TGFB1, and IGF2. It has been reported that in the absence of exogenously supplied feeders, a minor population of an ESC colony will differentiate into fibroblast-like cells in order to fulfill the role of feeders in establishing an *in vitro* niche. While the contents of basal media and supplemental growth factors are well defined, the profile of molecules secreted by either ESCs or feeder cells are not well understood. Thus the microenvironment, defined as the interface where ESCs within a colony exchange signals with each other as well as with feeder cells, remains to be further characterized.

While immunofluorescent labeling of ESC surface receptors can suggest a signal dependence on the cognate ligand for those receptors, it provides limited insight as to whether this pathway is autocrine or paracrine in nature. Bulk analysis of media conditioned by ESCs and feeders have been profiled with proteomic studies, but the distinct molecules within a heterogenous mixture cannot be traced back to the secretory cell of origin. Our platform uniquely enables the high thoroughput, quantitative profiling of candidate signaling proteins secreted by a single ESC or ESC-derivative. Furthermore, the ability to simultaneously assay a multiplex of secreted proteins can define and classify subpopulations of cells with distinct secretomes, several of which may be characteristic of pluripotent cells. Lastly, this novel diagnostic tool will also be useful in the field of regenerative medicine, particularly in studies related to the generation and application of induced pluripotent stem cell (iPSC) technology. The reprogramming of a somatic cell to a pluripotent cell is highly inefficient, and the mechanism underlying this process remains largely unknown. Probing the signaling aspects of a single cell undergoing the reprogramming process will yield novel insights into molecular mechanisms and thereby enhance the efficiency of the process. In terms of the clinical application of patient-specific iPSCs for cell replacement therapies, the safety of reprogrammed cells must be assessed prior to transplantation. Assaying for single cell secretomes anti-correlative to tumorigenic secretomes provides an additional parameter to ensure the safety of iPSCs used for regenerative medicine.

(558e) Neural Stem Cell 3D Neuronal Differentiation in Fluorinated Methacrylamide Chitosan Hydrogels

Hang Li[1], Asanka Wijekoon[2] and Nic D. Leipzig[1], (1)Department of Chemical and Biomolecular Engineering, The University of Akron, Akron, OH, (2)ARL BioPharma Inc., Oklahoma City, OK

Oxygen plays a vital role in the metabolism of living organisms and in tissue regrowth. Tissue engineering offers a potential solution to regenerate many tissues in the body including the central nervous system (CNS). Within the CNS, neural stem cells (NSCs) have been isolated that offer potential cellular treatments for neurodegenerative diseases, stroke, traumatic brain injury, and spinal cord injury. We recently invented a biocompatible photocrosslinkable chitosan hydrogel modified with perfluorocarbons (PFCs), which we call perfluorocarbon methacrylamide chitosan (MACF) that is able to uptake and release beneficial levels of oxygen to cells and tissues. The overall objective of this work is to demonstrate how MACFs can be utilized in 3D tissue engineered scaffolds to enhance cellularity and NSC differentiation.

In this study a series of novel, biocompatible, oxygen rich hydrogel scaffolds have been developed and synthesized by incorporating several PFCs to photopolymerizable methacrylamide chitosan hydrogels via Schiff base, nucleophilic substitution followed by radical polymerization to form hydrogels. The synthesized fluorinated methacrylamide chitosan (MACF) hydrogels were characterized by high resolution ^{19}F NMR and demonstrated the desired degree of fluorine substitution (35-45%). MACF oxygen adsorption and desorption studies demonstrated that MACF can be easily regenerated by reloading with oxygen once release is complete for at least three cycles with minimal change in properties. To study how oxygen rich hydrogels affect NSC survival and differentiation, NSCs were seeded and cultured in oxygenated and unoxygenated MACF hydrogels (MAC(Ali5)F with 5 aliphatic fluorine substitutions, MAC(Ar5F) with 5 aromatic fluorine substitutions and MAC(Ali15) with 15 aliphatic fluorine substitutions) and compared to nonfluorinated MAC hydrogels. After 8 days of differentiation and culture (in neuron specific medium), total cell number and neuronal differentiation was quantified. Results have revealed that MACFs support more cells than unfluorinated controls, especially in the interior of the scaffolds. Significant neuronal differentiation has been observed in all scaffolds, and we are currently studying neuronal extension and process formation.

(558f) Cytosolic Delivery of Reprogramming Factors with a Microfluidic Device

Nahyun Cho, MIT, Cambridge, MA

An efficient, robust cell delivery method for proteins has always been a daunting task. Current methods include electroporation, which creates pores in cell membranes to allow material to enter the cell, cell penetrating peptides, which modify the proteins to signal endocytosis, and polymers and liposomes, which encapsulate proteins and induce endocytosis. Each has its own flaws and are not considered universal delivery methods. For example, all three methods have been shown to be potentially toxic and the last two methods have been shown to be protein-specific.

By use of our microfluidic device, cells are forced through a constriction which creates pores on the membrane of the cell. This approach has many benefits, one of which is that our method is not material specific. Cells are suspended in solution with the protein of interest to deliver the protein. The second benefit of this method is the robustness to cell type. This microfluidic device works with established cell lines, immune cells, stem cells, as well as skin cells.

While this device is applicable to more than just protein delivery, this specific project focuses on the delivery of reprogramming factors, which are proteins, to skin cells to induce reprogramming into a pluripotent state. These induced pluripotent stem cells (iPS cells) have many benefits such as patient-specific tissues and the creation of immortalized model cell lines for diseases such as Parkinson's. The benefits of a fast and efficient method of creating iPS cells do not just extend to research, but also to potential clinical and tissue engineering applications as well.

(570a) Incremental Parameter Estimation and Ensemble Kinetic Modeling of Metabolic Networks

Gengjie Jia, Chemical and Bimolecular Engineering, Singapore MIT Alliance, Singapore, Singapore, Gregory N. Stephanopoulos, Chemical Engineering, Massachusetts Institute of Technology, Cambridge, MA and **Rudiyanto Gunawan**, Institute for Chemical and Bioengineering, ETH Zurich, Zurich, Switzerland

The estimation of unknown kinetic parameters from time-series measurements of biological molecules is a major bottleneck in the ODE model building process in systems biology and metabolic engineering (Chou and Voit, 2009). The majority of current estimation methods involve simultaneous (single-step) parameter identification, where model prediction errors are minimized over the entire parameter space. These methods often rely on global optimization methods, such as simulated annealing, genetic algorithms and other evolutionary approaches (Daisuke and Horton, 2006; Gonzalez, et al., 2007; Kikuchi, et al., 2003; Kimura, et al., 2005; Noman and Iba, 2007). The problem of obtaining the best-fit parameter estimates however is typically ill-posed due to issues related with data informativeness, problem formulation and parameter correlation, all of which contribute to the lack of complete parameter identifiability (Srinath and Gunawan, 2010). Not to mention, finding the global minimum of model residuals over highly multidimensional parameter space is challenging and can become prohibitively expensive to perform on a computer workstation, even for tens of parameters.

Here, we consider the modeling of cellular metabolism using the canonical power-law formalism, specifically the generalized mass action (GMA) systems (Savageau, 1969a; Savageau, 1969b). The power-law formalism has many advantages, which have been detailed elsewhere (Chou and Voit, 2009; Voit, 2000). Notably, power laws have a relatively simple structure that permits algebraic manipulation in the logarithmic scale, but nonetheless is capable of describing essentially any nonlinearity. Regulatory interactions among metabolites can also be described straightforwardly through the kinetic order parameters, establishing an equivalence between structural identification and parametric estimation. However, the number of parameters increases proportionally with the number of metabolites and fluxes, leading to a large-scale parameter identification problem, one where single-step estimation methods often struggle to converge.

As ODE integrations constitute a major part of the computational cost in the typical ODE parameter estimation (Voit and Almeida, 2004), alternative formulations have been proposed that avoid these integrations either completely (Tsai and Wang, 2005; Voit and Almeida, 2004) or partially (Jia, et al., 2011; Kimura, et al., 2005; Maki, et al., 2002). Particularly, computational cost could be significantly reduced by decomposing the estimation problem into two phases, starting with the calculation of dynamic reaction rates or fluxes from the slopes of concentration data, followed by the least square regressions of kinetic parameters (Bardow and Marquardt, 2004; Goel, et al., 2008; Marquardt, et al., 2006). In this case, the final parameter estimation is done one flux at a time, each involving only a handful of parameters and thus, the global minimum solution can be either computed analytically (for example, when using log-linear power-law flux functions) or determined efficiently. Moreover, as the first estimation phase (flux estimation) depends only on the assumption of the topology of the metabolic network, the flux estimates can subsequently be used to guide the selection of the most appropriate flux functions for the second phase or to detect inconsistencies in the assumed topology of the network separately from the flux equations (Goel, et al., 2008). However, the application of this method requires the number of metabolites to be equal to or larger than that of fluxes, so that the flux estimation can result in a unique solution. Since the reverse situation is more commonly encountered in the typical metabolic networks, a generalization of this incremental estimation approach is the main focus in this study.

As noted above, the new parameter estimation method in this work is built on the concept of incremental identification (Bardow and Marquardt, 2004; Marquardt, et al., 2006) or dynamical flux estimation (DFE) method (Goel, et al., 2008; Voit, et al., 2009), and designed to handle the degrees of freedom associated with the flux estimation from time-course concentration data. Specifically, two parameter estimation formulations are proposed with two different purposes. The first follows the most common formulation in which model prediction errors are minimized and a single "optimal" parameter set is produced. On the other hand, the second formulation is created toward the generation of an ensemble of parameter estimates and models, where each parameter combination will be able to fit the noisy concentration data equally well. The application of the first formulation to GMA models of a generic branched metabolic pathway and L. lactis glycolytic pathway (Voit, et al., 2006) demonstrates the significant improvement in the numerical efficiency offered by the proposed incremental approach, where the computational times were reduced by over two orders of magnitude in comparison with simultaneous (single-step) estimation methods. Meanwhile, in the second formulation, the ensemble of parameter estimates for the two GMA models above are constructed efficiently using an adaptive Monte Carlo and multiple ellipsoid sampling (Zamora-Sillero, et al., 2011). During this construction, the time-varying concentration simulations corresponding to the ensemble of models (or parameter estimates) are also produced. The ability to create such ensemble will necessitate the development of new framework for its use for metabolic engineering, which is a topic of our future work.

Bardow, A. and Marquardt, W. (2004) Incremental and simultaneous identification of reaction kinetics: methods and comparison, Chemical Engineering Science, 59, 2673-2684.

Chou, I.C. and Voit, E.O. (2009) Recent developments in parameter estimation and structure identification of biochemical and genomic systems, Math Biosci, 219, 57-83.

Daisuke, T. and Horton, P. (2006) Inference of scale-free networks from gene expression time series, Journal of bioinformatics and computational biology, 4, 503-514.

Goel, G., Chou, I.C. and Voit, E.O. (2008) System estimation from metabolic time-series data, Bioinformatics, 24, 2505-2511.

Gonzalez, O.R., et al. (2007) Parameter estimation using Simulated Annealing for S-system models of biochemical networks, Bioinformatics, 23, 480-48.

Jia, G., Stephanopoulos, G.N. and Gunawan, R. (2011) Parameter estimation of kinetic models from metabolic profiles: two-phase dynamic decoupling method, Bioinformatics, 27, 1964-1970.

Kikuchi, S., et al. (2003) Dynamic modeling of genetic networks using genetic algorithm and S-system, Bioinformatics, 19, 643-650.

Kimura, S., et al. (2005) Inference of S-system models of genetic networks using a cooperative coevolutionary algorithm, Bioinformatics, 21, 1154-1163.

Maki, Y., et al. (2002) Inference of genetic network using the expression profile time course data of mouse P19 cells, Genome Informatics, 13, 382-383.

Marquardt, W., Brendel, M. and Bonvin, D. (2006) Incremental identification of kinetic models for homogeneous reaction systems, Chemical Engineering Science, 61, 5404-5420.

Noman, N. and Iba, H. (2007) Inferring gene regulatory networks using differential evolution with local search heuristics, IEEE/ACM transactions on computational biology and bioinformatics / IEEE, ACM, 4, 634-647.

Savageau, M.A. (1969a) Biochemical systems analysis. I. Some mathematical properties of the rate law for the component enzymatic reactions, J Theor Biol, 25, 365-369.

Savageau, M.A. (1969b) Biochemical systems analysis. II. The steady-state solutions for an n-pool system using a power-law approximation, J Theor Biol, 25, 370-379.

Srinath, S. and Gunawan, R. (2010) Parameter identifiability of power-law biochemical system models, J Biotechnol, 149, 132-140.

Tsai, K.Y. and Wang, F.S. (2005) Evolutionary optimization with data collocation for reverse engineering of biological networks, Bioinformatics, 21, 1180-1188.

Voit, E.O. (2000) Computational analysis of biochemical systems : a practical guide for biochemists and molecular biologists. Cambridge University Press, New York.

Voit, E.O. and Almeida, J. (2004) Decoupling dynamical systems for pathway identification from metabolic profiles, Bioinformatics, 20, 1670-1681.

Voit, E.O., et al. (2006) Regulation of glycolysis in Lactococcus lactis: an unfinished systems biological case study, Syst Biol (Stevenage), 153, 286-298.

Voit, E.O., et al. (2009) Estimation of metabolic pathway systems from different data sources, IET Syst Biol, 3, 513-522.

Zamora-Sillero, E., et al. (2011) Efficient characterization of high-dimensional parameter spaces for systems biology, BMC Syst Biol, 5, 142.

(570b) Rational Design of 13C-Labeling Experiments for Metabolic Flux Analysis Using Elementary Metabolite Unit-Basis Vectors (EMU-BV)

Maciek R. Antoniewicz, Department of Chemical and Biomolecular Engineering, University of Delaware, Newark, DE and Scott B. Crown, Chemical Engineering, University of Delaware, Newark

^{13}C-Tracer selection is an important step in conducting ^{13}C-metabolic flux analysis (^{13}C-MFA) studies; however, current methods are restricted to trial-and-error approaches, which commonly focus on an arbitrary subset of the tracer design space. To effectively select optimal tracers, there is a pressing need for new rational approaches. Towards this end, we have introduced the concept of elementary metabolite unit basis vectors (EMU-BV). We demonstrate that any metabolite in a network model can be described as a linear combination of EMU-BVs, where the corresponding coefficients indicate the fractional contribution of the EMU-BV to the product metabolite. The strength of our approach is the decoupling of the substrate labeling, i.e. the EMU-BVs, from the dependence on free fluxes, i.e. the coefficients. We demonstrate that flux observability depends inherently on the number of independent EMU-BVs and the sensitivities of coefficients with respect to free fluxes.

The EMU-BV methodology was applied to design tracer experiments in microbial and mammalian systems of interest. First, we applied it to a model of HEK-293 metabolism to select optimal tracers for elucidating two key fluxes, the oxidative pentose phosphate pathway (oxPPP) and pyruvate carboxylase (PC) fluxes. Through efficient grouping of coefficient sensitivities, simple tracer selection rules were derived for high-resolution of the fluxes. The approach resulted in a significant reduction of the number of possible tracers and the feasible tracers were evaluated using numerical simulations. Two optimal, novel tracers were identified, specifically [2,3,4,5,6-13C]glucose for elucidating oxPPP flux and [3,4-13C]glucose for elucidating PC flux. The novel tracers improved confidence intervals of the estimated fluxes by 3- to 10-fold compared to currently used tracers. Secondly, we investigated flux branch points in microbial metabolism. In particular, we developed a high-precision method to estimate the oxidative PP pathway, Entner Doudoroff, and glycolysis fluxes with a single tracer experiment. Using glycolysis intermediates as measurements, candidate tracers were identified and numerical simulations narrowed the candidate list to a handful of optimal tracers. Experimental determination of the split ratios was conducted in two model organisms, E. coli and S. cerevisiae, to validate the modeling predictions. In summary, our work highlights the power of rational selection of isotopic tracers and provides a roadmap for high resolution flux measurements.

(570c) Thermodynamics-Based Flux-Balance Analysis: Incorporation of Thermodynamic and Metabolomic Data Into Genome-Scale Constraint-Based Models

Joshua J. Hamilton and Jennifer L. Reed, Department of Chemical and Biological Engineering, University of Wisconsin-Madison, Madison, WI

Genome-scale metabolic reconstructions provide a concise, mathematical representation of cellular metabolism, and can be used as tools to probe a variety of biological questions via constraint-based methods. These methods have been successfully used to predict cellular phenotypes, guide metabolic engineering, and provide context to high-throughput ('-omics') data. The predictive accuracy of constraint-based methods relies on an accurate representation of the underlying biochemical network, including both the network stoichiometry and the quantitative assignment of reaction directions. Existing approaches to quantitatively assign reaction directionality rely on group contribution theory [1] to calculate thermodynamically feasible directions of individual reactions based on a standard metabolite concentration [2], or to enforce thermodynamic feasibility via the imposition of additional reversibility constraints for reactions identified as qualitatively reversible [3].

We have used a thermodynamic-based flux balance analysis approach, TFBA, to ensure thermodynamically feasible directions for all reactions in a genome-scale metabolic network. We used group contribution theory estimates and errors of Gibbs free energies of formation for all metabolites, and calculate Gibbs free energies of reaction in response to metabolite concentrations. Metabolite concentrations were allowed to vary freely, or were constrained based on experimental measurements. We compared TFBA predictions for growth phenotypes of *Escherichia coli* single-gene knockout mutants to both experimental data and to existing flux balance analysis predictions. We also examined the sensitivity of predicted growth rate to uncertainties in intracellular metabolite concentrations and Gibbs free energies of formation. We subsequently identified a set of metabolites for which precise concentration and thermodynamic data are necessary for accurate *in silico*predictions. These results emphasize the need for precise metabolomic data and experimental measurements of thermodynamic properties of key metabolites.

References

[1] Jankowski, et al. "Group contribution method for thermodynamic analysis of complex metabolic networks." *Biophysical Journal*2008. 95:1487-1499.

[2] Fleming, et al. "Quantitative assignment of reaction directionality in constraint-based models of metabolism: Application to Escherichia coli." *Biophysical Chemistry*2009. 145:47-56.

[3] Henry, et al. "Thermodynamics-Based Metabolic Flux Analysis." *Biophysical Journal* 2007. 92:1792-1805.

(570d) Discriminating Significant From Insignificant Model Parameters: The Case of a Dynamic CHO Cell Model

Hana Sheikh, Department of Chemical and Biological Engineering, Tufts University, Medford, MA, Kyongbum Lee, Chemical and Biological Engineering, Tufts University, Medford, MA and Christos Georgakis, Systems Research Institute and Chemical and Biological Engineering, Tufts University, Medford, MA

Kinetic modeling for metabolic networks, formulated as a set of ordinary differential equations for intracellular species concentrations, provides the ability to simulate the dynamic behavior of cellular metabolism. Such models aim to predict cellular response to various external stimuli, allowing an investigator to develop a detailed fundamental understanding of the phenomena studied. Investigators often choose to include more than the necessary model details rather than risk the error of including less than necessary details. However, this increases the number of parameters that need to be identified from experimental data and introduces a substantial mathematical limitation in the identification of the important parameters. As more details are added to the model, the increased number of parameters implies an increased amount of experimental data. Most importantly, it is not clear whether all the model details and the corresponding parameters are necessary for a desired set of model predictions. The present paper presents a computationally efficient methodology to identify the model parameters that are significant for the model predictions and thus distinguish them from the insignificant ones.

The proposed approach is inspired by the classic Design of Experiments (DOE) techniques, performed *in silico* using a preliminary model. We start by defining the possible range of each of the unknown model parameters and designing a set of experiments that are simulated through the preliminary model. Utilizing analysis of variance (ANOVA) and response surface model (RSM) linear regression tools we develop a simplified nonlinear meta-model in which only the significant parameters are retained. We applied this methodology to a dynamic model of Chinese hamster ovary (CHO) cell metabolism [Nolan et al. 2010]. This model, comprising 51 parameters and 34 reaction fluxes, was able to provide a reliable preliminary prediction of the effects of fed-batch process variables such as temperature shift, specific productivity, and nutrient concentrations. A D-optimal design of experiments was used to sample the parameters across their ranges, and a RSM was obtained with antibody flux as the output. Investigating linear, linearly interactive, and quadratic RSMs, we efficiently eliminated approximately 90% of the terms as being insignificant, shedding light on the importance of each of the 51 original model parameters in the predictions of the metabolic model.

(570e) Metabolic Flux-Based Modularity Using Shortest Retroactive Distances

Gautham V. Sridharan[1], Michael Yi[1], Soha Hassoun[2] and Kyongbum Lee[1], (1)Chemical and Biological Engineering, Tufts University, Medford, MA, (2)Computer Science, Tufts University, Medford, MA

Modularity analysis offers a route to better understand the organization of cellular biochemical networks as well as to derive practically useful, simplified models of these complex systems. While there is general agreement regarding the qualitative properties of a biochemical module, there is no clear consensus on the quantitative criteria that may be used to systematically derive these modules. Recently, we proposed a round trip distance metric, termed Shortest Retroactive Distance (ShReD), to characterize the cyclical interaction between any two reactions in a biochemical network and to group together network components that mutually influence each other [1]. Applied to a liver metabolic network, the ShReD partitions found hierarchically arranged modules, many of which contained reactions that were cyclically related by the shared consumption and production of cofactors. One limitation of this analysis was that it relied on uniform edge weights, implicitly assuming that all the reactions in the network are equally engaged. In this work, we update the ShReD-base partitioning algorithm with a novel edge-weighting scheme that enables the use of metabolic flux data for metabolic state-dependent modularity analysis.

In a reaction-centric graph model of a metabolic network, a quantitative measure of the interaction between a pair of connected reaction nodes can be obtained from the flux of the intermediary metabolite. We define the weight of an edge between a connected pair of reaction nodes as the *inverse of the fraction of the intermediate metabolite production flux that is directed towards the destination node*. This definition is consistent with the intuitive notion that a large metabolite flux from one reaction to another corresponds to a strong interaction, and thus is represented by a relatively short edge, whereas a weak interaction is represented by a long edge. To examine the impact of the flux weights, we applied the modified partitioning algorithm to flux data describing adipocyte differentiation and enzyme inhibition [2]. Our results indicate that the metabolic state of the adipocyte significantly impacts the modular assignment of two key upstream reactions in fatty acid synthesis and glycerogenesis, pyruvate carboxylase (PCX), and phosphoenolpyruvate carboxykinase (PEPCK), which connect carbohydrate metabolism to lipogenesis. Our analysis also identifies several robust reaction pairs that consistently partition together into the same module regardless of metabolic state. Examples include reaction pairs that couple the pentose phosphate shunt and palmitate synthesis through the production and consumption of NADPH. Lastly, we show that the modular organization of adipocyte metabolism is relatively stable with respect to the inhibition of an enzyme compared to a major physiological change such as cellular differentiation.

[1] Sridharan GV, Hassoun S, Lee K (2011) Identification of Biochemical Network Modules Based on Shortest Retroactive Distances. PLoS Comput Biol 7(11): e1002262. doi:10.1371/journal.pcbi.1002262

[2] Si, Yaguang, Hai Shi, and Kyongbum Lee. "Impact of perturbed pyruvate metabolism on adipocyte triglyceride accumulation." *Metabolic engineering* 11, no. 6 (November 2009): 382-90.

(570f) Analysis of Critical Transitions in a Model of Human Endotoxemia

Jeremy D. Scheff, Biomedical Engineering, Rutgers University, Piscataway, NJ, Steve E. Calvano, Department of Surgery , UMDNJ-Robert Wood Johnson Medical School, New Brunswick, NJ and Ioannis P. Androulakis, Department of Chemical and Biochemical Engineering, Rutgers, The State University of New Jersey, Piscataway, NJ

Outcomes from inflammation-linked diseases are determined not only by the severity of injury or the virulence of a pathogen, but also by the body's endogenous inflammatory response. An appropriate inflammatory response leads to healing and recovery, but either too strong or too weak of a response can lead to complications. This is particularly relevant in critically ill patients, whose inflammatory networks are at heightened states of activity as their bodies attempts to recover from their precarious states. A critically ill patient can undergo transitions to different states, for instance representing a return to healthy homeostasis, the development of a concomitant disease, or death, depending on the interactions between their immune response and their illness. Identification and analysis of "tipping points" between states could not only allow for more informed therapy of inflammation-linked diseases, but could also have significant diagnostic and prognostic implications (An, Nieman et al. 2012). Discrete tipping points as a stimulus is gradually increased have been observed in a variety of computational, experimental and clinical environments (An, Nieman et al. 2012). In the context of inflammation, feedforward loops between damage and inflammatory mediators can become activated in response to severe injury, leading to bistability and the potential for abrupt switching between states when a threshold is crossed. Therefore, the identification and analysis of inflammation-driven tipping points is extremely important, particularly if a forthcoming tipping point can be predicted ahead of time to allow for more focused intervention. We approached this issue based on a model of human endotoxemia (Scheff, Mavroudis et al. 2011) to investigate critical transitions within the context of inflammation and study how tipping points emerge; under what conditions the resolution of inflammation and the restoration of homeostasis are possible; how critical transitions between states can be predicted ahead of time; and how circadian homeostatic control systems factor into the transition between states. We evaluated the hypothesis that a gradually changing disturbance provokes a critical transition between steady states of the inflammatory response, and furthermore that this transition can be predicted by stability analysis and that there are underlying circadian patterns in these responses.

Human endotoxemia, an experimental model of systemic inflammation, consists of the administration of low doses of endotoxin (lipopolysaccharides, LPS) to healthy human volunteers. This provokes physiological changes that, in part, mimic those occurring in inflammation-linked diseases such as sepsis and trauma. We previously designed a mathematical model of human endotoxemia (Scheff, Mavroudis et al. 2011) which incorporates a feedforward loop in the inflammatory response that leads to bistability, with one steady state representing healthy homeostasis with low levels of inflammatory mediators, and the other steady state exhibiting heightened levels of inflammatory mediators that persist even when LPS is removed. In this work, we studied the response to chronic LPS exposure, mimicking a slowly growing bacterial infection. Under these conditions, the system initially remains in an intermediate transition state, where state variables are very gradually evolving, until there is an abrupt change and the heightened inflammatory state is reached. To assess the stability of this model as a function of time, the Jacobian matrix was numerically estimated. Given the eigenvalues of a linearizable system, it is possible to determine its stability, has been used an indicator of an impending transition in ecological models (Lade and Gross 2012). All simulations were performed multiple times throughout the day to evaluate the impact of circadian rhythms on the system.

We identified that, after the system becomes unstable, there is a delay before the trajectory begins to significantly move towards the heightened inflammatory state, as shown in Figure 1. This represents a critical time window where identification of an impeding state transition is possible, but intervention still allows the system to return to homeostasis. Additionally, there are circadian dependences on both the time of onset of this time window as well as its size, illustrating how homeostatic control systems can play a role in the response to stress. These results indicate how critical information about a system, information which may be of significant translational value, can be buried within the dynamic behavior of that system.

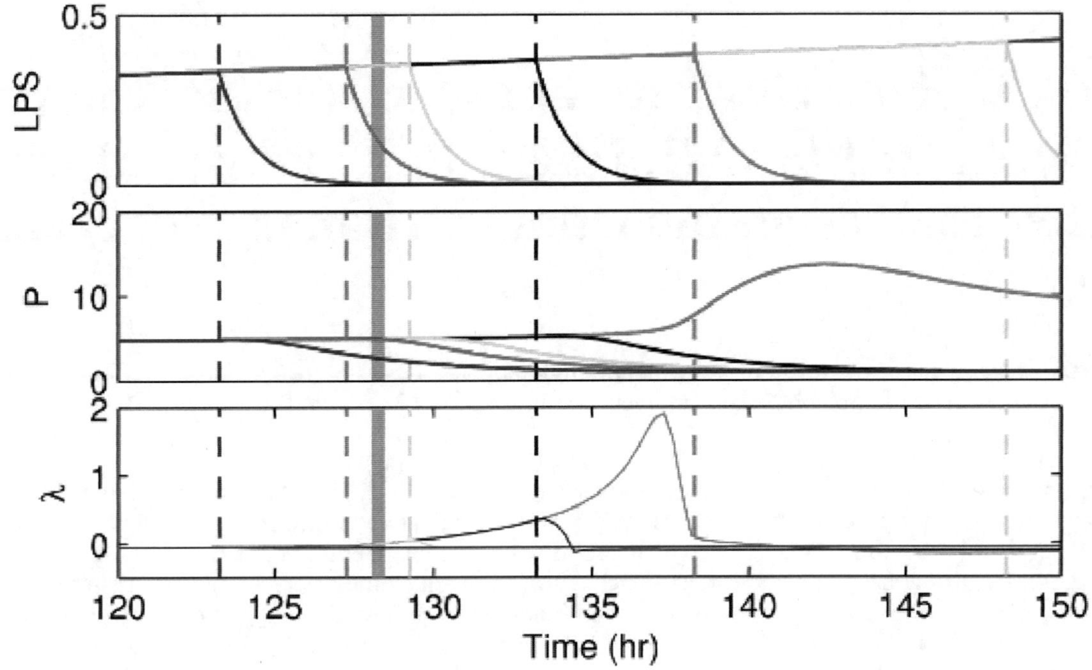

***Figure 1**: LPS exposure is initiated at t=0hr. The three panels respectively show two different model variables (LPS and P, pro-inflammatory signaling) and the largest eigenvalue of the Jacobian (λ) as functions of time. Each thin solid curve represents a separate simulation, each with identical beginnings, but with LPS removed at different time points, denoted by the vertical dashed lines. The thick vertical grey bars mark the time at which the maximum eigenvalue of the Jacobian matrix became positive, signaling the impending transition between steady states.*

This type of analysis is a first step towards identifying how the dynamical components of the inflammatory network can be leveraged to predict forthcoming tipping points as well as to more broadly understand system stability properties. As progress is made towards addressing the issues raised here, we will grow ever-closer to the ideal of building personalizable models based on clinically available biomarkers. A model calibrated on an individual patient could provide a personalized trajectory that could be analyzed for its stability, allowing for more focused care when a critical transition is imminent.

References

- An, G., G. Nieman, et al. (2012). "Computational and systems biology in trauma and sepsis: current state and future perspectives." Int J Burn Trauma 2(1): 1-10.
- An, G., G. Nieman, et al. (2012). "Toward Computational Identification of Multiscale "Tipping Points" in Acute Inflammation and Multiple Organ Failure." Ann Biomed Eng.
- Lade, S. J. and T. Gross (2012). "Early warning signals for critical transitions: a generalized modeling approach." PLoS Comput Biol 8(2): e1002360.
- Scheff, J. D., P. D. Mavroudis, et al. (2011). "Modeling autonomic regulation of cardiac function and heart rate variability in human endotoxemia." Physiol Genomics 43(16): 951-64.

(570g) A Global Sensitivity Approach for the Analysis of Intracellular PI3K/AKT Signaling Pathway During Definitive Endoderm Induction of Human Embryonic Stem Cells

Shibin Mathew, Department of Chemical and Petroleum Engineering, University of Pittsburgh, Pittsburgh, PA and Ipsita Banerjee, Chemical and Petroleum Engg, University of Pittsburgh, Pittsburgh, PA

Introduction: Differentiation of human embryonic stem cells (hESCs) to the definitive endoderm lineage is an important step towards important functional cell types of the pancreas and liver. *In vitro*differentiation efforts are directed towards the activation of cellular signaling pathways which are active during embryonic development in vertebrates. As an important step in the chemical induction of definitive endoderm, modulation of the growth factor mediated PI3K/AKT pathway has been implicated in the control of the outcome of other pathways by making early fate decisions between the maintenance of pluripotency and initiation of differentiation [1]. Specifically, PI3K/AKT pathway suppresses the differentiation signals and sustains the undifferentiated state, while inhibition of the pathway results in up-regulation of Smad and Wnt signaling pathways [2]. Mathematical models can help in the formulation of informative experiments to identify the precise role of these interactions and the conditions under which the cell commits to a particular fate. However, by nature, stem cell systems exhibit intricate signaling pathways; and with the added uncertainty in physical parameters, mechanistic models are either incomplete or difficult to build. One possible solution to construct predictable models is to perform sensitivity analysis on models validated in other cellular contexts and identify regions of interest relating to an experimentally observed output. Given the multivariate nature of the problem, such a sensitivity analysis has to be global in approach to capture the nonlinear interactions between the system parameters. The goal of our study was the use of global sensitivity analysis to design experiments to analyze the role of PI3K/AKT pathway following the induction of hESCs towards definitive endoderm.

Methods: IGF mediated activation of the PI3K/AKT pathway was analyzed using the model developed by Sedaghat *et al* [3]. The model encompasses receptor binding dynamics followed by activation of intracellular signaling cascade resulting in the phosphorylation of a key node in the pathway, i.e. AKT. The activation of Akt has been correlated to various phenotypic outcomes of the cell related to survival, differentiation, stemness, migration etc. We based our global sensitivity analysis on random sampling high dimensional model representation (RS-HDMR) developed by Rabitz *et al* [4]. The technique constructs a meta-model of the original signaling pathway and by using orthogonal function decomposition techniques, evaluates the well-known Sobol' sensitivity indices of the system input parameters to the model output. The advantage of using RS-HDMR lies in the reduction in number of random samples required ($\sim 10^3$ - 10^4) to estimate the sensitivity indices as compared to computationally expensive brute-force Monte Carlo methods ($> 10^6$). Through our analysis, we identified key rate parameters to which the phospho-Akt concentrations are sensitive to, in the presence and absence of feedback mechanisms. Our analysis revealed interesting behaviors of phopho-Akt and its functional dependence on the various system parameters. This knowledge was used to design an *in vitro* strategy to inhibit the PI3K/AKT pathway and study the phospho-Akt protein dynamics during the initial stages of differentiating hESCs.

Results: According to our sensitivity analysis, PTEN (phosphatase and tensin homolog), which is a negative regulator of the PI3K/AKT pathway and parameters associated with receptor activation were the most sensitive parameters in the absence of feedback. However, with feedback, the *in silico* phopho-Akt population segregated into high phospho-Akt and low phospho-Akt regions. RS-HDMR in each of these regions revealed differences in sensitive parameters, particularly to negative feedback loops which inhibit the ability of tyrosine phosphorylated IRS (Insulin receptor substrate) to complex with PI3K and activate downstream Akt. Interestingly, even after PI3K inhibition, our experiments showed increase in phospho-Akt in the initial stages of differentiation and Smad activation and it hints towards a possible cross-talk with other pathways like Smad and Wnt involved in differentiation which maintain higher phopho-Akt levels by modulation of the likely sensitive parameters.

Conclusion: RS-HDMR with polynomial approximation was successful in identifying the most sensitive parameters in the insulin signaling pathway and compared well to the brute-force Monte Carlo estimates. Although different variants of RS-HDMR exist, our analysis proves that even simple orthonormal polynomial approximations can give accurate ranking of the sensitive parameters and often, such a ranking is what we seek to prioritize experimental resources to reducing the uncertainty of the sensitive inputs. Our integrated mathematical and experimental approach was successful in explaining the important parameters affecting the PI3K/AKT pathway in differentiating hESCs.

References

1. McLean AB, D'Amour KA, Jones KL, Krishnamoorthy M, Kulik MJ, Reynolds DM, Sheppard AM, Liu H, Xu Y, Baetge EE: **Activin A efficiently specifies definitive endoderm from human embryonic stem cells only when phosphatidylinositol 3-kinase signaling is suppressed.** *Stem Cells* 2007, **25:**29-38.

2. Singh AM, Reynolds D, Cliff T, Ohtsuka S, Mattheyses AL, Sun Y, Menendez L, Kulik M, Dalton S: **Signaling Network Crosstalk in Human Pluripotent Cells: A Smad2/3-Regulated Switch that Controls the Balance between Self-Renewal and Differentiation.** *Cell Stem Cell* 2012, **10:**312-326.

3. Sedaghat AR, Sherman A, Quon MJ: **A mathematical model of metabolic insulin signaling pathways.** *American Journal of Physiology-Endocrinology and Metabolism* 2002, **283:**E1084-E1101.

4. Li G, Rosenthal C, Rabitz H: **High dimensional model representations.** *The Journal of Physical Chemistry A* 2001, **105:**7765-7777.

(570h) A Mathematical Model of Tumor-Induced Bone Disease Based On the Vicious Cycle Concept

Junhwan Jeon, Deptartment of Chemical and Biomolecular Engineering, Vanderbilt University, Nashville, TN, Scott A. Guelcher, Department of Chemical and Biomolecular Engineering, Vanderbilt University, Nashville, TN, Julie A. Sterling, Cancer Biology, Vanderbilt University, Nashville, TN and Peter T. Cummings, Center of Nanophase Materials Sciences, Oak Ridge National Laboratory, Oak Ridge, TN

Tumor cells grow in specific metastatic sites because of molecular interactions that occur between tumor cells and their microenvironment. Several tumor cells such as breast and prostate cancers frequently metastasize to bone where they release factors that cause bone destruction thwarting normal bone remodeling. During the bone destruction, growth factors are released from bone matrix that stimulate further growth of the tumor cells and subsequently more bone resorbing factors are released from the tumor cells. In order to understand underlying mechanism of this "vicious cycle", a mathematical model has been developed based on a recent mathematical model of bone remodeling by Komarova et al. [1]. Our model describes the dynamics of cell population and cell-cell interactions among osteoclasts, osteoblasts, and tumor cells within a single bone modeling unit where the net effects of local factors on the rates of cell populations have been approximated using power law relations. Tumor burden and bone mass change calculated from the model are compared to values obtained from experiments in which MDA-MB-231 cells, a human cell line that preferentially metastasizes to bone, were injected directly into the tibiae of mice. It is found that the model captures the autocatalytic mechanism of bone destruction observed experimentally, demonstrating at least qualitative validation of the model.

[1] Komarova SV, Smith RJ, Dixon SJ, Sims SM, Wahl LM. Mathematical model predicts a critical role for osteoclast autocrine regulation in the control of bone remodeling. BONE 2003;33: 206-215.

(615a) The Kinetics of Fibrin Clot Formation On Surfaces Immobilized with Cell and Fibronectin Binding Domains

Anand Ramanathan and Nancy Wangechi Karuri, Chemical and Biological Engineering, Illinois Institute of Technology, Chicago, IL

Fibronectin (FN) is essential for the formation of a fibrin clot, an extracellular matrix that arrests bleeding. In the clot, FN binds fibrin and provides a scaffold for wound healing to occur. Our laboratory had previously shown that the ratio between the amount of a FN and cell binding domains, III_{1-2} and III_{9-10} respectively, immobilized on a surface influences the amount of FN in the ECM of cells. We investigated the effect of surfaces with different amounts of III_{1-2} and III_{9-10} on the kinetics of fibrin clot formation in a cell free system. Affinity purified glutathione tagged III_{1-2} and III_{9-10} were coated on polystyrene dishes. ELISA assays were used to identify the limits of protein loading on the surface and to quantify the levels of the individual proteins in mixtures. Fibrinogen was added to surfaces immobilized with III_{1-2} and $III_{9-10,}$ fibrin clot formation was proteolytically activated by the addition of thrombin and followed through dynamic light scattering. The kinetics of clot formation were studied in the presence and absence of full-length FN and with or without the covalent cross-linker, Factor XIIIA. We found that the rate of fibrin clot assembly was dependent on the ratio of III_{1-2} to III_{9-10} and higher rates were obtained when the ratio of III_{1-2} to III_{9-10} was high. Moreover, the effect of the III_{1-2} and III_{9-10} coated surfaces on the rate of clot formation was significantly enhanced by FN and factor XIIIA. These results in a cell free system coupled together with our previous findings provide new avenues for the design of materials that can be used to accelerate wound healing.

(615b) Injectable Thermo-Sensitive Hydrogel As an Adjuvant: In Vivo Modulation of Dendritic Cells for Cancer Vaccine

Kye-Il Joo, Liang Xiao, Yarong Liu and Pin Wang, Chemical Engineering and Materials Science, University of Southern California, Los Angeles, CA

Development of an effective vaccine for cancer involves eliciting a potent antitumor immune response by introduction of tumor antigens to trigger systemic immunity against the tumor. Dendritic Cells (DCs) are the most powerful antigen-presenting cells capable of initiating adaptive immune response by stimulating both T cells and B cells and therefore become one of the major target cells for the cancer vaccine development. Immunization by adoptive transfer of autologous DCs that are loaded with antigens *in vitro* is, however, labor-intensive and requires significant regulatory concerns. In addition, most transplanted DCs remain non-functional, and few home to the lymph nodes where the subsequent activation of T-cell immunity is induced. More cost-effective and efficient protocols involve delivering antigen to host DCs *in vivo* by direct injection of vaccine vectors to program DCs in situ. However, *in vivo* DC targeting approach is limited by low specificity and poor efficacy of antigen uptake to DCs. Thus, the ability to efficiently deliver antigens to host DCs *in vivo* is a crucial step towards generating a potent antitumor immunity.

In this study, we hypothesized that creating *in vivo* microenvironment to recruit and proliferate DCs, which mimics natural surroundings of infection in the body, may potentially provide a practical mean to enhance the uptake of antigens to host DCs and precisely control the timing of DC-trafficking and activation *in situ*. Therefore, we investigated the development of the functionalized biomaterials incorporating inflammatory cytokines as infection-mimicking microenvironment for recruiting and housing host DCs *in vivo* prior to introduce tumor antigens at the DC-enriched site. This approach involves the design and use of thermo-sensitive mPEG-PLGA hydrogel capable of solution-gel transition in the response to temperature, which facilitates drug loading and exhibits sustained release of the drug. The implantation of the hydrogel loaded with granulocyte-macrophage colony-stimulating factor (GM-CSF), which has been identified as a potent stimulator of DC recruitment and proliferation, could provide the subcutaneous microenvironment where host DCs could be recruited and proliferated by the sustained and localized release of GM-CSF at the hydrogel-injected site. Analysis of in vivo DC recruitment showed that the number of recruited DCs was peaked at around 7 to 10 days post-injection of GM-CSF hydrogel. The migration assay indicated that the recruited DCs were able to migrate to lymph node where T-cell response was triggered. Furthermore, our lymphocyte analysis suggested that the significantly improved antigen-specific immune responses were observed when immunogens (i.e. antigens) were administrated at the site of GM-CSF-hydrogel implanted at 7 days post-injection, compared to empty hydrogel or without hydrogel. In this presentation, this hydrogel adjuvant system with many different immunogens including lentiviral vector and liposome/protein will be discussed in more detail.

(615c) Control of Adult Stem Cells with Polyurethane Gel Matrix

Rajesh Krishnan[1], Sandeep Nalluri[2] and **Debanjan Sarkar**[1,2], (1)Biomedical Engineering, University at Buffalo, The State University of New York, Buffalo, NY, (2)Chemical & Biological Engineering, University at Buffalo, The State University of New York, Buffalo, NY

Control of stem cell fate in tissue engineering application is important for effective and functional tissue regeneration. Extracellular matrix provides structural and functional cues in a highly synchronized fashion to stem cells for controlling the cellular processes and cell fate. Synthetic matrices are designed to induce similar interactions between stem cell and microenvironment which can effectively guide stem cells with spatial and temporal control. Biocompatible and biodegradable polyurethanes with segmental composition can be designed as gel with nanostructured phases due to the presence of 'soft' and 'hard' segments. By tuning the structure and composition of polyurethanes, the inter- and intra-segmental interactions can be modulated to control the polyurethanes into gel like structure. We have designed polyurethanes which forms gel through physicochemical interactions. Specifically, degradable peptide-based polyurethanes were synthesized and their interactions with adult stem cells were examined in 2-D substrate and 3-D environment. We examined the effect of polyurethane gel structure by varying the soft and hard segment. Soft segment of the polyurethanes were altered by using polyethylene glycol of different molecular weight and hard segment was designed from aliphatic diisocyanate with dipeptide chain extender. Our results shows that the composition of polyurethane chains with segments induce specific gel structure. Thus, by controlling the segmental architecture of polyurethanes, morphology of the gels were controlled to provide signals to stem cells. Interaction of bone marrow derived mesenchymal stem cells (MSCs) with the polyurethane gels shows specific dependence of the polyurethane structure and gel morphology. MSCs interacted with the different segments of polyurethane gels to present distinct cellular structure and behavior. Specifically, cell adhesion, aggregation, cellular structure and morphology were influenced by the polyurethane gel matrix. Interaction of MSCs with polyurethane reveals that segmental composition and gel morphology of polyurethanes influences the cell function and behavior which can control the fate of MSCs, particularly the differentiation of the stem cells. Biosynthetic activities of MSCs within the polyurethane gels were modulated to produce extracellular material leading to formation of tissue like structure. This study indicates that molecular engineering of polyurethane gel provide exquisite control of stem cell behavior. Thus, controlling cell-material interactions with these matrices can modulate tissue regeneration.

(615d) Correlating Effects of Gel Microstructural Features with Specific Differentiation Patterning of Mouse Embryonic Stem Cells

Keith Task[1], Antonio D'Amore[2], Satish Singh[1], Maria Jaramillo[3], Prashant Kumta[2] and Ipsita Banerjee[4], (1)Chemical Engineering, University of Pittsburgh, Pittsburgh, PA, (2)Bioengineering, University of Pittsburgh, Pittsburgh, PA, (3)Depatment of Bioengineering, University of Pittsburgh, Pittsburgh, PA, (4)Chemical and Petroleum Engg, University of Pittsburgh, Pittsburgh, PA

Embryonic stem cells (ESC) have the potential to be used in many applications due to their ability to differentiate into cells of any of the three germ layers (ectoderm, endoderm, and mesoderm). There are several different routes by which ESC can be guided to differentiate into specific germ layers, including genetic manipulation, chemical cues, and mechanical stimuli. With respect to the latter, it has been observed that substrate stiffness plays a significant role in governing specific phenotype differentiation.

In our previous study we investigated the effect of mechanical stiffness of fibrin on ESC differentiation. Our results indicated that while ectoderm and mesoderm germ layers displayed weak response to the change in fibrin substrate stiffness in the chosen range (2 – 250KPa), endoderm markers were strongly responsive, with softer substrates up-regulating endoderm specific markers. Although cells respond differently to substrates of varying stiffness, it was also observed that gels with the same stiffness but fabricated under different conditions lead to different differentiation patterning. Therefore, this macroscopic property is perhaps not what the cell truly experiences; modification of fabrication conditions changes the gel microstructure, which in turn affects the stiffness, and these micro-characteristics directly interact with the cells. However, it is not clear what specific microstructural features are the most influential in inducing cellular differentiation, and correspondingly, how they affect ESC behavior.

In the current work, we investigate the effect of microstructural features of fibrin gels on the differentiation of mouse ESC. Twelve different fibrin gels were fabricated by varying the fibrinogen concentration and fibrinogen to thrombin crosslinking ratio. The fibrin gels were used to induce ESC differentiation employing both 2D and 3D cultures. After the differentiation protocol the ESC were analyzed for phenotypic commitment by performing qRT-PCR for the specific germ layer markers. Each of the 12 different fibrin gels was imaged with scanning electron microscopy. Microstructural features of each of these gels were quantified using an image analysis tool for the characterization of fibrous scaffolds [1]. Specific features which were characterized include fiber diameter, node density, fiber length, and pore size. A comparison of these attributes along with principal component analysis led to a subset of features which vary most across gel conditions. Gel microstructural features were correlated with the ESC differentiation patterning using regression analysis. The gels are heterogeneous in nature, and therefore a bootstrapping technique was used to account for the variability of the system. Interestingly, only a subset of genes shows a significant relationship with microstructural features, and that only a few of these features significantly affect differentiation, including fibrin pore size distribution and node density. These results are also compared to the correlation of gel stiffness, determined by both by rheological measurements and atomic force microscopy, to phenotype commitment. This analysis reveals the sensitivity of cellular phenotype commitment and differentiation patterning to each of the microstructural features. Moreover, such information can be used to help guide the design of scaffolds with specific properties for tissue engineering applications. Results of these studies will be presented and discussed.

[1] A. D'Amore *et al.* Biomaterials, 2010, Vol. 31, 5345-5354.

(615e) A Novel Application of Nanofiber Scaffolds for Gene Therapy

Mandula Borjigin[1], Bryan Strouse[1], Pawel Bialk[1], Chris Eskridge[1], Rohina Niamat[1], Jingwei Xie[2] and Eric Kmiec[1], (1)Chemistry, Delaware State University, Dover, DE, (2)Marshall Insititute for Interdisciplinary Research, Marshall University, Huntington, WV

Electrospun nanofiber scaffolds have been used in cell growth and tissue engineering. Here we expand the application of nanofiber scaffolds to the genetic technique known as gene editing. Gene editing is a process by which single base mutations can be corrected, in the context of the chromosome, using a short piece of single stranded DNA in the form of an oligonucleotides (ssODN). The ssODNs are introduced into the cell to correct the mutant nucleotide using the inherent pathways of DNA replication and repair. This technique is being developed rapidly for use in the treatment of inherited diseases such as Sickle Cell Disease (SCD) and Phenylketonuria (PKU). The survival and proliferation of the cells bearing modified genes however appear to be impeded by a phenomenon known as reduced proliferation phenotype (RPP); this is the key barrier to practical implementation of gene editing in the clinical setting. To overcome RPP, we have utilized nanofiber scaffolds as templates on which modified cells were allowed to recover, grow and expand after gene editing. Here, we present evidence that HCT116-19, bearing an integrated, mutated eGFP gene and corrected by gene editing, proliferate on polylysine or fibronectin coated polycaprolactone (PCL) nanofiber scaffolds. In contrast, cells from the same reaction protocol plated on both regular plain surfaces and polylysine (or fibronectin) coated plate surfaces do not proliferate. Therefore, growing genetically modified (edited) cells on electrospun nanofiber scaffolds promotes the reversal of the Reduced Proliferation Phenotype and increases the potential of gene editing as an *ex vivo* gene therapy application.

(615f) Topography Mediated Regulation of HER-2 Expression in Breast Cancer Cells

Amita Daverey, Austin Mytty and **Srivatsan Kidambi**, Dept of Chemical & Biomolecular Engineering, University of Nebraska-Lincoln, LINCOLN, NE

The breast tumor microenvironment is made up of heterogenous mixture of pores, ridges and collagen fibers with well defined topographical features. Although, significant progress has been achieved toward elucidating the biochemical and molecular mechanisms that underlie breast cancer progression, quantitative characterization of the associated mechanical/topographical properties and their role in breast tumor progression remains largely incomplete. Therefore, the aim of the study is to investigate the effect of topography on the adhesion and biology of breast cancer cells in *in* vitro cultures. Breast cancer cells were cultured on different topographies designed on polyelectrolyte multilayer (PEM) coated polydimethylsiloxane (PDMS) surfaces. We demonstrate that microtopography affects the cell adhesion and distribution depending on the topography on the PDMS surfaces. Further, we report that surface topography down-regulates the HER-2 gene and protein levels in breast cancer cells when compared to the tissue culture polystyrene surfaces (TCPS) control. Results from this study indicate that topography of different patterns modulates morphology of cells, their distribution and expression of HER-2 in breast cancer cells. The observations made by this study provide rational basis for understanding the native topography of breast in order to design bio-mimetic tissue culture substrate for better understanding of the disease.

(615g) The Effect of Length and Terminal Group of Poly(ethylene glycol)-Terminated Self-Assembled Monolayers (SAMs) On Dendritic Cell (DC) Maturation and Function

Christina Yacoob[1], Jung Park[2], Bingbing Sun[1] and Hong Shen[1], (1)Department of Chemical Engineering, University of Washington, Seattle, WA, (2)Department of Neuroscience and Behavioral Biology, Emory University, Atlanta, GA

Dendritic cells (DCs) are the most potent antigen presenting cells (APCs). They are the primary target for inducing effective immune responses or tolerance. *Ex-vivo* induced DCs have been used for immunotherapy of cancers. DCs are present around implanted devices though their functions remain to be elucidated. Oligomers or polymers of ethylene glycol (EG) have been grafted onto surfaces of various biomaterials for controlling protein adsorptions and increasing the biocompatibility of devices. In this study, immature DCs were exposed to poly(ethylene glycol) (PEG)-terminated self-assembled monolayers (SAMs). OH-terminated PEG SAMs activated immature DCs and upregulated surface molecules characteristic of mature DCs at a higher degree than COOH-terminated ones. Longer OH-terminated PEG SAM encouraged the maturation of DCs at a higher magnitude than shorter ones. Interestingly, none of SAMs induced cytokines characteristic of either inflammatory or tolerogenic DCs. Additionally, we demonstrated that OH-terminated PEG SAM activated DCs mainly through DAP-12 or Fc-gamma – associated surface receptors-dependent pathway. Our results give insights into biomaterial design for inducing activation or tolerance of DC-based immunity. Since DAP-12 or Fc-gamma – associated surface receptors-dependent signaling pathway is involved in many cell types, including macrophages, our results also suggest that PEG-terminated SAMs can potentially activate other cell types in a length and functional end-dependent manner.

(615h) An Interactive Engineered Protein Hydrogel: Controlling and Responding to Neurite Growth

Kyle J. Lampe and Sarah C. Heilshorn, Materials Science and Engineering, Stanford University, Stanford, CA

Neural regeneration within the central nervous system (CNS) is a critical unmet challenge as brain and CNS disorders continue to be the leading cause of disability nationwide. Common tissue engineering goals seek to customize cell-biomaterial interactions and guide cell behavior. Here we have developed a material that is both cell instructive and cell responsive, creating a dynamic interplay between neural cells and their engineered extracellular matrix (ECM). Using recombinant protein technology, we engineered a family of elastin-like protein hydrogels with multiple independently tunable properties. With these materials we can investigate individual and synergistic effects of elastic modulus, degradation rate, and cell adhesivity on cell behavior in a tunable microenvironment. Unlike natural matrices, the concentration of adhesive ligands such as the fibronectin-derived sequence RGD can be precisely controlled without altering the mechanical properties of the hydrogel. When dorsal root ganglion (DRG) neurons were cultured in 3D in these hydrogels, a cell-adhesive RGD density of 1.9×10^7 ligands/μm^3 promoted neuron-specific growth and more than doubled the rate of neurite extension compared to hydrogels without the adhesive sequence. Crosslinking density was tuned to create scaffolds with elastic moduli ranging from 0.5-2.1 kPa with constant adhesive ligand densities. The most compliant gels led to the greatest outgrowth from encapsulated DRGs with neurites extending over 1800 μm by day 7. In contrast, the stiffest gels permitted far fewer extensions and limited outgrowth to a maximum of 600 μm over the same time frame.

To render the materials cell-responsive, we designed the engineered proteins to contain a second set of bioactive sequences that specifically respond to changes in cell phenotype. Neural stem cells (NSCs) undergoing differentiation may change their production of the protease urokinase plasminogen activator (uPA), which has previously been found at the growth cones of extending neurites. By incorporating cell-mediated degradable subunits into the elastin-like proteins, we are able to mimic the natural remodeling of the ECM. We engineered multiple uPA target sites with different degradation kinetics into the elastin-like protein to allow neural cell-mediated control of the scaffold degradation dynamics. DRGs encapsulated in uPA degradable scaffolds extended neurites at a faster rate than in hydrogels of similar RGD density and initial stiffness without the uPA degradable sequence. This cell-responsive strategy was also used to enhance the functionality of the material by controlling delivery of multiple molecules with distinct release kinetics.

These crosslinked scaffolds are useful for directing the growth and differentiation of multiple cell types including clinically relevant NSCs. The tunable scaffolds are responsive to neuronal cells, which may be able to specifically self-modulate the release of multiple bioactive factors while undergoing differentiation. This work demonstrates the versatility and responsiveness of our modularly-designed protein hydrogels for neural cell culture and encourages continued development as a biomaterial tissue construct for treating spinal cord injury.

(621a) Development of Three-Dimensional Lung Multicellular Spheroids in Air and Liquid Interface Culture for the Evaluation of Anti-Cancer Therapeutics

Samantha A. Meenach[1,2], Alexandra N. Tsoras[1], Ronald C. McGarry[3], Heidi M. Mansour[2], J. Zach Hilt[1] and Kimberly W. Anderson[1], (1)Chemical and Materials Engineering, University of Kentucky, Lexington, KY, (2)Pharmaceutical Sciences - Drug Development Division, University of Kentucky, Lexington, KY, (3)Radiation Medicine, University of Kentucky, Lexington, KY

Applying *in vitro* outcomes to *in vivo* applications has limitations because conventional two-dimensional (2D) cell culture does not recreate a physiologically representative model for cells. This work investigated a three-dimensional (3D) cell culture technique to model lung tumors *in vitro*. A 3D lung cancer model was created by applying collagen to a cell culture Transwell, which allows for nutrient transfer through the collagen. Two lung cancer cells lines (H358, a bronchioalveolar carcinoma and A549, a lung adenocarcinoma) were seeded on the collagen. The non-adhesive collagen caused the cells to be more inclined to attach to one another rather than its surface allowing for the formation of multicellular spheroids (MCS). To better mimic the environment for lung cancer specifically, an air-interface culture (AIC) condition was created. For AIC conditions, the cell media on the apical side of the Transwell was removed, and the basolateral side with media provided effective nutrient transport to the MCS while still exposing the cells to air. 2D versus 3D cell behavior in response to paclitaxel and aerosol particles containing paclitaxel was evaluated using the viability of the cells.

Through the use of brightfield and fluorescence microscopy imaging, the AIC model yielded viable MCS at sizes similar to MCS formed in LCC conditions (100 to 150 μm in diameter). With the optimized 3D model, LCC cells were exposed to paclitaxel in media, and paclitaxel-loaded dry powder aerosol particles were delivered to AIC cells through direct application with an insufflator. This alternative delivery method using dry powder particles was used since paclitaxel could not be delivered through media as in LCC conditions. The particles containing paclitaxel were comprised of a PEGylated phospholipid excipient mixture which encapsulated the drug. Using viability analysis, it was shown that both applications of paclitaxel showed variance in efficacy when comparing 2D and 3D culture conditions (where the IC_{50} values for paclitaxel were higher for 3D compared to 2D). Transepithelial electrical resistance of A549 cell monolayers was evaluated before and after particle delivery to illustrate that the particle application does not affect the permeability of the cells, which infers that this form of drug therapy will not affect the permeability of lung tissue. Overall, a much more representative *in vitro* model has been developed that is expected to be an improved predictor of efficacy of alternative drug delivery methods such as direct pulmonary delivery for lung cancer, which could lead to more efficient drug therapies for lung cancer patients.

(621b) Treatment Tests of Breast Cancer in Mice Using a Vascular-Targeted Enzyme Prodrug Therapy

Brent D. Van Rite, Chemical, Biological and Materials Engineering, University of Oklahoma , Norman , OK, Vassilios I. Sikavitsas, Chemical, Biological, and Materials Engineering, University of Oklahoma, Norman, OK, Mohamad Cherry, Hematology-Oncology , University of Oklahoma Health Sciences Center , Oklahoma City, OK, Carla Kurkjian, Hematology-Oncology, University of Oklahoma Health Sciences Center, Oklahoma City, OK and **Roger G. Harrison**, Chemical, Biological and Materials Engineering, University of Oklahoma, Norman, OK

Metastatic breast cancer is usually not curable. The median survival of women with metastatic breast cancer is approximately two years after documentation of metastasis. Treatment for most patients with metastatic breast cancer is considered palliative in nature, although durable complete remissions have been achieved in a small number of patients (< 5%). Some disease response has been demonstrated with various chemotherapeutic agents; however, these agents are almost always associated with undesirable side effects such as nausea, vomiting, hair loss, cognitive dysfunction, and fatigue. Recent research has centered on therapies that are directed only towards malignant cells.

This work is on a novel enzyme prodrug therapy using a fusion protein containing annexin V to target only tumor vascular endothelial cells and cancer cells, reducing unwanted toxicity to healthy tissue. The L-methioninase enzyme will convert the inert prodrug selenomethionine to toxic methylselenol and also deplete the cancer cells of methionine necessary for protein synthesis and continued growth. The L-methioninase enzyme is linked to human annexin V protein in a fusion protein. This therapy takes advantage of the fact that phosphatidylserine (PS), an anionic phospholipid, is exposed almost exclusively on the outer membrane of cancer and endothelial cells of the tumor vasculature and that annexin V binds specifically to PS.

Recombinant technology was used to express and purify the methioninase-annexin V fusion protein. During purification, a $(His)_6$ tag was cleaved from the fusion protein, and endotoxin was removed. Pharmacokinetic testing in mice revealed complete clearance of the fusion protein from the bloodstream to occur within 12 hours following intraperitoneal (i.p.) injection. The enzyme prodrug system was tested in SCID (severe combined immune deficient) female mice with implanted with MDA-MB-231 human breast tumors and was found to significantly inhibit tumor growth compared to three control groups of mice (untreated, fusion protein only, and selenomethionine only). The fusion protein and the prodrug were delivered by i.p. injection. The MDA-MB-231 cancer cells had been transfected with green fluorescent protein (GFP) to follow the growth of the tumors using a small animal fluorescence imaging system. Results will also be reported of a separate study of this enzyme prodrug therapy using mice fed a diet deficient in methionine but supplemented with homocysteine.

(621c) Quorum-Sensing Salmonella Selectively Trigger Protein Expression Upon Colonization within Tumors

Charles A. Swofford, Adam T. St. Jean and Neil S. Forbes, Chemical Engineering, University of Massachusetts, Amherst, MA

Engineered *salmonella* possess unique capabilities that make them ideal drug delivery vectors for tumors. Current cancer chemotherapeutic regimes have limited efficacy due to therapeutic resistance, which decreases the success rate for treatment of late-stage cancer and metastases. This therapeutic resistance can be explained by a number of factors including 1) intra-tumoral transport limitations, 2) reduced susceptibility of quiescent cancer cells, and 3) protein pumps that expel therapeutics from cells. Bacteria are capable of overcoming these therapeutic limitations because they preferentially accumulate in tumors, actively penetrate tumor tissue, and can be engineered to produce anti-cancer drugs *in situ*. Due to systemic toxicity, however, bacteria that constitutively express an anti-cancer drug would not be an ideal alternative to standard chemotherapy. Therefore, we have engineered *salmonella* to secrete anti-cancer drugs after tumoral colonization has occurred through the use of the *lux* quorum sensing system. Quorum sensing is the ability for bacteria to change their gene expression based on changes in population density. Because *salmonella* preferentially accumulate in tumors, a quorum sensing genetic circuit, coupled with an anti-cancer agent, creates a robust expression switch that only turns on in bacterial tumor colonies. This unique bacterial drug delivery vector will allow for targeted delivery of a powerful anti-cancer drug to therapeutically resistant tumors and metastases while minimizing the risk of systemic toxicity. *Salmonella* transformed with the quorum-sensing system induce expression upon reaching a threshold bacterial density of 10^8 cfu/mL (OD = 0.5). Furthermore, in a tumor-on-a-chip device, quorum-sensing bacteria induce expression once colonies have formed within the tissue. These colonies have formed deep within the therapeutically resistant regions of tumors where most standard chemotherapeutic drugs cannot reach due to diffusion limitations. Additionally, 4T1 murine tumor models injected with the quorum-sensing system show co-localization of GFP and *salmonella* within tumor tissue while healthy tissue shows low-counts of *salmonella* without any GFP expression. Finally, quorum-sensing *salmonella* secrete therapeutic proteins that reduce tumor burden and enhance mouse survival. The culmination of this data demonstrates that *salmonella* integrated with a quorum-sensing expression vector can sense colonization within tumors and induce protein expression once a critical density has been reached. Bacteria transformed with a quorum-sensing system will also form colonies deeper into the therapeutically resistant regions of tumors allowing for treatment of tumor tissue previously unreachable by standard chemotherapeutic regiments. Overall, this system creates a robust on/off switch that will allow for targeted bacterial drug delivery to tumors and metastases while minimizing the risk of potentially harmful drugs reaching healthy tissue.

(621d) Inferring Tumor-Immune Interaction Networks Via Unbiased Secretome Profiling

Yogesh Kulkarni[1], **Kisheon Alexander**[1], Yueting Wu[1] and David J. Klinke[1,2,3], (1)Chemical Engineering, West Virginia University, Morgantown, WV, (2)Microbiology, Immunology and Cell Biology, West Virginia University, Morgantown, WV, (3)Mary Babb Randolph Cancer Center, School of Medicine, West Virginia University, Morgantown, WV

Despite the variability of the disease, cancer cells exhibit a set of common traits. Tumor-escape from the cytotoxic action of the immune system is an emerging hallmark of cancer but the interaction between the tumor cell and the host is complex. We employ experimental procedures in limited length and time scales to elucidate the mechanisms by which tumors escape immunosurveillance, and predict the effect on physiology. On understanding the cross-talk between cancer and immune cells, more comprehensive targeted therapies may be developed. This study aims to compare the biochemical signals secreted in vitro by malignant cells to normal human cells. Two human mammary epithelial cancer cell lines, BT474 (HER2+/ER+) and SKBR3 (HER2+/ER-) and one normal human mammary epithelial cell line, 184A1 were investigated. A proteomics workflow that includes two-dimensional gel electrophoreses, MALDI-TOF mass spectrometry and peptide mass finger-printing were used to analyze the secretome. Protein interaction networks and canonical pathways were extracted from the Ingenuity Knowledge Base to elucidate the role of tumor-derived secreted protein on immune cell function. The presence of tumor-derived exosomes was confirmed by SEM, and initial results suggest them as extra-cellular protein conveyers. Future studies will concentrate on inhibiting tumor cells from evading elimination. Proteomics provides a less biased approach to characterize the biochemical signals that malignant cells use to skew immunity and to better understand cell-to-cell communication in breast cancer.

(621e) Design of A Biomimetic Nanostructured Tissue Engineered Bone Model for in Vitro Breast Cancer Metastasis Study

Mian Wang, Department of Mechanical & Aerospace Engineering, George Washington University, Washington, DC, Sidney Fu, George Washington University, Washington, DC and Lijie Grace Zhang, Department of Mechanical and Aerospace Engineering and Institute for Biomedical Engineering, The George Washington University, Washington, DC

Breast cancer is the leading cause of cancer-related death in American women, accounting for more than 200,000 new cases and 40,000 deaths annually. Majority of advanced breast cancer patients will eventually develop bone metastasis. However, the cure and treatment of metastasis breast cancer is still very challenging because of insufficient knowledge about underlying metastasis mechanisms. Traditional tumor metastasis models (*in vivo* and *in vitro*) contain inherent limitations with regards to controllability, reproducibility, and flexibility. To solve the above problems, this research designs an easy-to-use and biomimetic 3D nanostructured bone model via natural chitosan and various hydroxyapatites (HA) for breast cancer metastasis study. 3D porous chitosan-HA scaffolds with different crystalline, size and concentration of HA (specifically, 10% and 20% nanocrystalline HA, microcrystalline HA and amorphous HA) were fabricated via a lyophilization procedure. Human MDA-MB231 metastasis breast cancer cells were seeded at 50000 cells/scaffold. After 24h cell culture, the cell adhesion and growth result reveals that concentration, size and crystallinity of HA have a significant impact on metastasis breast cancer cell behavior. 10% nanocrystalline HA can significantly increase breast cancer cell attachment when compared to all of other HA. Since natural bone matrix is made of rich nanocrystalline HA, our synthesized nanocrystalline HA in chitosan can create a more bone-like environment for breast cancer cell metastasis when compared to microcrystalline and amorphous HA. Furthermore, 20% nanocrystalline HA in chitosan shows the similar increasing trend for cancer cell attachment when compared to respective 20% microcrystalline HA in chitosan. In addition, the result reveals that lower concentration of HA (10%) can promote more breast cancer cell growth than higher concentration of HA (20%) in chitosan. Based on the result, 10 % nanocrystalline HA-chitosan scaffold can significantly improve more breast cancer cell growth than other size, concentration and crystallinity of HA chitosan scaffold, thus make it promising for further breast cancer cell metastasis study.

(621f) Mechanistic Investigation of the TRAIL-Sensitization Activity of Mitoxantrone

David J. Taylor, Chemical Engineering, Arizona State University, Tempe, AZ, Taraka Sai Pavan Grandhi, Biomedical Engineering, Arizona State University, Tempe, AZ, Arul Jayaraman, Department of Chemical Engineering, Texas A&M University, College Station, TX and Kaushal Rege, Chemical Engineering, Biomedical Engineering, Arizona State University, Tempe, AZ

In the United States, prostate cancer is the most commonly diagnosed cancer in men, and results in the second largest number of cancer-related fatalities (in men). Similarly, breast cancer is the most commonly diagnosed cancer in women, and results in the second largest number of fatalities (in women). Pancreatic cancer ranks as the fourth most fatal malignancy in both, men and women; the five-year survival rate is less than 5%. Recent studies from our laboratory resulted in the identification of an FDA-approved drug, mitoxantrone, for sensitizing prostate, pancreatic, and breast cancer cells to Tumor necrosis factor Related Apoptosis Inducing Ligand (TRAIL)-induced death. Here we report an investigation into the sub-cellular mechanisms by which mitoxantrone sensitizes resistant cancer cells to TRAIL. In the investigation, kinetic studies were performed to determine the period of time over which sensitization occurred. The concentrations of apoptosis specific proteins were also measured to identify changes following mitoxantrone treatments. Additionally, the kinetics of mitochondrial depolarization, generation of reactive oxygen species (ROS), and apoptosis induction were measured to determine the factors that contribute to the activity of the drug. Finally, nanoparticles were investigated as effective methods of delivering these potent combination treatments to cancer cells, with an eye toward enhancing efficacy and specificity.

(621g) Differential Effects of Chemokines and Growth Factors On Tumor Cell Migration Using a 3D Microfluidic in Vitro Model

Beum Jun Kim[1], Pimkhuan Hannanta-anan[1], Yoon Soo Kim[2], Michelle Chau[1], Melody A. Swartz[3] and Mingming Wu[1], (1)Biological and Environmental Engineering, Cornell University, Ithaca, NY, (2)Biomedical Engineering, Cornell University, Ithaca, NY, (3)Institute for Bioengineering, École Polytechnique Fédérale de Lausanne (EPFL), Lausanne, Switzerland

Tumor cell chemotaxis within the interstitial space is an important transport mechanism in cancer metastasis. Despite its clinical importance, the molecular mechanism which tumor cells use to migrate in chemokine or growth factor gradients is largely unknown. Using a recently developed 3D microfluidic gradient generating device, we studied tumor cell migration in well-defined and steady linear chemical gradients. We used chemokines, SDF-1α and CCL19, and a growth factor, EGF, which are three well-known cytokines present in tumor microenvironments, to generate chemical gradients; we also used a malignant breast tumor cell line, MDA-MB-231, embedded in type I collagen for 3D cell cultures. Our experimental results demonstrate (1) The major motility type in our 3D microfluidic culture is amoeboid motility: tumor cells also exhibit plasticity and heterogeneity in motility; (2) Both EGF and CCL19 promote tumor cell motility only: the speed increases synergistically when 0.25 nM EGF and 111 nM/mm SDF-1α are applied; (3) Breast tumor cells are chemotactic towards high concentrations of SDF-1α with enhanced motility: this chemotactic behavior follows a ligand-receptor binding kinetics (11~222 nM/mm); (4) The chemotactic sensitivity to a SDF-1α gradient of 111 nM/mm diminishes when either 0.25 or 8.33 nM EGF is supplemented. These results will eventually help us better understand the physical and chemical factors in tumor cell migration, in the context of microenvironmental interactions.

(621h) Analysis of Myc-Driven Metabolic Reprogramming in B-Cells by Isotopically Nonstationary 13C Flux Analysis

Taylor A. Murphy[1], Chi Dang[2] and **Jamey D. Young**[1], (1)Chemical and Biomolecular Engineering, Vanderbilt University, Nashville, TN, (2)Abramson Cancer Center

The transcription factor c-Myc is an oncoprotein whose expression is deregulated in over 30% of all cancers. A significant role of c-Myc (i.e., Myc) is its control of cell cycling and apoptosis, but it also exerts strong influence on central metabolism. Targeting metabolic phenotypes induced by Myc activation may offer a potential therapeutic strategy to selectively inhibit the proliferation of Myc-driven cancers. We have conducted ^{13}C metabolic flux analysis (^{13}C-MFA) studies on P493-6 B-cells to identify metabolic pathway alterations associated with Myc overexpression. These cells have a tetracycline-repressible genetic construct that enables three distinct c-Myc expression levels: High (oncogenic), Low (endogenous), and None. Using GC-MS profiling techniques on isotopically labeled protein and RNA samples, along with measurements of extracellular metabolite uptake and excretion fluxes, we have created comprehensive flux maps of the central carbon metabolism of P493-6 cells under both High and Low Myc conditions. These flux maps have revealed several key features of Myc metabolic regulation.

We observed that glycolysis is highly upregulated in both conditions, with lactate-to-glucose (L/G) ratios of 1.54 and 1.79 for High and Low Myc, respectively. However, 25% of the pyruvate synthesized via glycolysis in the High Myc condition entered the TCA cycle as compared to only 15% in the Low Myc cells. Most TCA cycle and amphibolic mitochondrial pathways exhibited 4-fold flux increases in High Myc cells, in contrast to modest increases in glucose uptake and lactate excretion. This was largely driven by increased catabolism of glutamine and other amino acids. The oxidative pentose phosphate pathway exhibited minimal activity under both High and Low Myc conditions. Oxygen uptake was significantly increased in High Myc cells as compared to Low Myc, confirming the increase in mitochondrial activity detected by MFA. We are currently conducting studies to quantify how hypoxia modulates these metabolic phenotypes as well as examining siRNA knockdowns of specific enzymes that are activated in High Myc cells.

(726a) Cyclic Strain Versus Endothelial Cell Presence On MSC Osteogenesis

Mariah S. Hahn, Department of Chemical Engineering, Texas A&M University, College Station, TX

Mesenchymal stem cells (MSCs) are increasingly recognized as a viable cell source for bone regeneration applications. In addition to soluble factors, cyclic stretch has been shown to have a profound effect on mesenchymal stem cell (MSC) osteogenesis. Similarly, endothelial cell (EC) presence has been demonstrated to enhance MSC osteogenic differentiation. However, potential synergistic interactions between mechanical stimulation and EC presence remain to be elucidated. The aim of the present manuscript was therefore to examine the simultaneous influence of cyclic stretch and EC paracrine signaling on MSC osteogenesis in the context of scaffolds with "osteogenic" moduli.

To accomplish this, 10T½ multipotent stem cells were encapsulated in poly(ethylene glycol) diacrylate [PEGDA] hydrogels with moduli within the "osteogenic" range. Half of the constructs were fabricated with a luminal EC layer. EC^+ and EC^- constructs were then subjected to continuous cyclic stretch. Following 10 days of culture, both EC^+ and EC^- constructs were associated with significantly elevated levels of chondrogenic transcription factor sox9 relative to initial (day 0) expression levels. By day 22 of culture, however, sox9 levels in both the EC^+ and EC^- constructs had returned to or fallen below day 0 levels. In contrast, osteocalcin expression was significantly higher in Day 22 EC^+ constructs relative to EC^- constructs and relative to day 0. Similarly, osteopontin and alkaline phosphatase levels were elevated in Day 22 EC^+ constructs relative to EC^- constructs. Cumulatively, the present results suggest that EC paracrine signaling enhances MSC osteogenesis in the presence of cyclic stretch. In addition, the observed transition from chondrogenic to osteogenic protein expression associated with the EC^+ constructs would be consistent with the neovascularization-dependent transition from cartilage matrix to osteoid matrix associated with endochondral bone formation.

(726b) Co-Electrospun Scaffolds with Gradients in Fiber Alignment and Chemistry for Regeneration of the Ligament-Bone Interface

Satyavrata Samavedi, Prudvi Gaddam, Abby Whittington and Aaron Goldstein, Chemical Engineering, Virginia Tech, Blacksburg, VA

Co-electrospun scaffolds with gradients in fiber alignment and chemistry for regeneration of the ligament-bone interface

Introduction: Grafts currently used for the repair of ligament injuries integrate poorly with bone due to a significant mismatch of mechanical and chemical properties between graft and bone. This mismatch stems from the grafts' inability to regenerate gradients of mineral content, collagen fiber alignment and cell phenotype that exist at natural ligament-bone transitions. We envision that an electrospun scaffold, possessing gradients of architecture, mechanical and biochemical properties, will help establish spatial gradients of cell phenotype in bone marrow stromal cells (BMSCs) and thus aid the regeneration and osseointegration of injured ligaments. As a step towards the larger goal of regenerating complex tissue transitions, we describe herein the fabrication of a scaffold possessing gradients of mineral content and fiber alignment, obtained by co-electrospinning two polymer solutions, with different chemistries, from offset spinnerets onto a dual-drum collector. The dual-drum collector used in this study allows the creation of graded scaffolds with transition regions at physiologically relevant length-scales.

Materials and Methods: Solutions of poly-lactic-co-glycolic acid (PLGA) doped with an amorphous calcium phosphate (ACP) phase and polycaprolactone (PCL) were co-electrospun from offset spinnerets onto a custom-designed slowly rotating dual-drum collector (**Figure 1a**). Post-electrospinning, the drums were stretched apart to result in strain-induced fiber alignment in the gap region between the drums. This fabrication procedure resulted in the formation of randomly oriented ACP doped-PLGA fibers on one drum, aligned PCL fibers in the gap region between the drums and a transition region consisting of both types of fibers at the interface. Angular orientation of the fibers from the three regions was characterized from micrographs of the scaffold. Finally, rat BMSC alignment, cell metabolic activity, deposition of Collagen-I (Col-I) and phenotypic makers of cell differentiation were investigated on all regions of the graded scaffolds.

Results and Discussion: Co-electrospinning onto the custom-designed collector resulted in the formation of a graded scaffold consisting of three different regions: a randomly oriented region possessing fibers that were electrospun from one spinneret (**Figure 1b**), an aligned region possessing fibers that were electrospun from the other spinneret (**Figure 1d**) and a transition region possessing a mixture of both types of fibers (**Figure 1 c**). After 1 and 3 days of culture, rat BMSCs were found to be aligned in the aligned region of the scaffold (**Figure 1e**), and randomly oriented on the other regions. Cells were also found to be metabolically active on all regions of the scaffolds after 1 and 7 days of culture. Finally, a gradient of cell phenotype and collagen-I (Col-I) orientation was observed on the graded scaffolds.

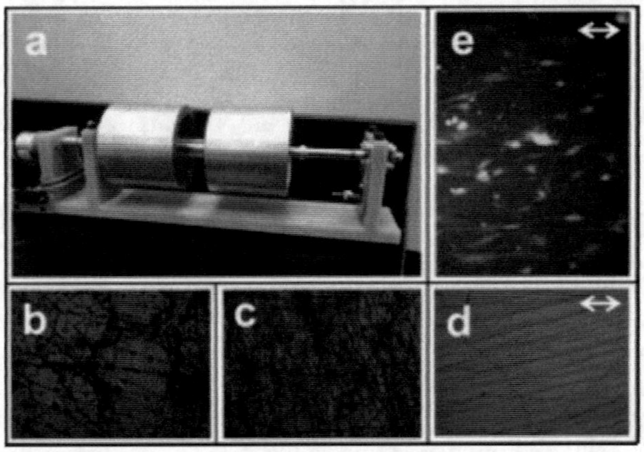

Figure 1:

(a) Slowly-rotating dual-drum set-up with gap region in between the drums.

(b), (c) and (d): Microscopy images of samples collected from a scaffold fabricated with two PCL solutions from offset spinnerets (proof-of concept). White arrow in (d) indicates general direction of fiber alignment.

e) Rat bone marrow stromal cells seeded on the aligned PCL section scaffold and imaged with Calein AM after 1 day of seeding. White arrow indicates general direction of cell alignment.

Conclusions: This study demonstrates that scaffolds with complex architectures and gradients in mechano-chemical properties at physiologically relevant length-scales, can be fabricated by co-electrospinning appropriate polymer solutions from offset spinnerets onto a specially designed collector. The study also indicates that these scaffolds have the potential to be used for the regeneration of graded tissue transitions such as those present at the ligament-bone interface.

(726c) Engineered Arterial Mimics (EAMs) to Quantify Smooth Muscle Cell Contribution to Atherosclerosis

William Herrick, Chemical Engineering, University of Massachusetts Amherst, Amherst, MA and Shelly R. Peyton, Chemical Engineering, University of Massachussetts, Amherst, Amherst, MA

For over a century, coronary heart disease (CHD) has been the leading cause of death in the Western world, surpassing all forms of cancer combined. The primary cause of CHD, atherosclerosis, is an extensively studied, yet incompletely understood disease. We hypothesize that one reason this disease is poorly understood, as well as why so few drug targets discovered *in vitro* are successful *in vivo*, is that the conditions in which cells are cultured and studied *in vitro* do not capture the complex *in vivo* microenvironment. To address this issue, we have created Engineered Arterial Mimics (EAMs) from PEG-PC gels, a hydrogel formed from a PEGDMA (poly(ethylene) glycol dimethacrylate) backbone with zwitterionic phosphorylcholine (PC) pendant groups, and which include integrin-binding extracellular matrix (ECM) proteins found in healthy and atherosclerotic arteries.

PEG-PC gels offer key advantages over PEG hydrogels. First, the Young's modulus can be tuned over a wider range by allowing gel formation at very low PEGDMA concentrations (~2 kPa, 5 kPa, 50 kPa, 200 kPa and 500 kPa at 0.5, 1, 3, 6 and 10% PEGDMA and 20% PC). Second, they do not become opaque at higher Young's moduli, and finally, the zwitterionic PC group makes these gels more hydrophilic than PEGDMA alone and they can, therefore, more efficiently block non-specific protein adsorption. We are exploiting these features to create EAMs that represent the chemical and physical changes that occur during the progression of atherosclerosis. We have EAMs that consists of collagen III, collagen IV and laminin (70/15/15 weight % at 10 ug/cm^2 total protein), which are predominantly found in healthy arteries, and EAMs that consist of fibronectin and collagen I (50/50 weight % at 10 ug/cm^2 total protein), which dominate the atherosclerotic artery. In order to capture the reported mechanical heterogeneity of diseased arteries, EAMs are made over a wide range of Young's moduli: ~5, 50 and 200 kPa (1, 3, 6% PEGDMA and 20% PC, respectively).

Preliminary migration experiments on EAMs varying only integrin-binding proteins found that human aortic smooth muscle cells (SMCs) migrate faster on the diseased ECM than on the healthy ECM (20 µm/hr vs. 15 µm/hr), indicating that the adhesive ligands available to the cells are able to tip the 'phenotypic balance.' We are currently performing migration experiments on PEG-PC gels to inform us of how the integrin-binding and stiffness cues integrate to affect this particular phenotype indicator. Preliminary results indicate that calponin, an early SMC differentiation marker, has a biphasic dependence on substrate stiffness (relative expression on 5, 50 and 200 kPa PEG-PC gels: 1.0, 1.6, 1.2), whereas phosphorylation of the 20 kDa myosin light chain (pMLC
calponin (relative expression on 5,50 and 200 kPa PEG-PC gels: 1.0, 0.69, 0.94). We quantified SMC proliferation over the course of a week on the EAMs in either serum-free DMEM, DMEM containing smooth muscle growth supplement (SMGS; 5% FBS, 2 ng/mL basic fibroblast growth factor (bFGF), 0.5 ng/mL epidermal growth factor, 5 ng/mL heparin, 5 ug/mL insulin) or a defined media condition (serum-free, 25 ng/mL platelet-derived growth factor, and 10 ng/mL bFGF). We found that proliferation increases with stiffness in the no-serum and defined media conditions, proliferation is higher on the diseased EAM than the healthy EAM, and proliferation in the SMGS condition is high on all EAM and stiffness conditions. The latter observation suggests that serum and/or various mitogens can override the effect of ECM ligands and stiffness on phenotype maintenance.

We are continuing to investigate the effect of substrate stiffness and ECM ligands on proliferation, migration and SMC marker expression. In particular, we are interested in the interplay between SMC mechanosensing (integrin binding, cytoskeleton assembly and acto-myosin contractility), growth factor sensitivity and SMC phenotype. If we can better understand how these factors are related in a system, which closely resembles the *in vivo* environment, there is a greater potential to make discoveries that will lead to novel drug targets and treatments.

(726d) Development of Biomimetic Environments with Appropriate Chemical and Mechanical Cues for Cells in Bioengineered Vascular Grafts

Mao-Shih Liang, Chemical and Biological Engineering, State University of New York at Buffalo, Amherst, NY, Maxwell T. Koobatian, Physiology, State University of New York at Buffalo, Amherst, NY, Daniel D Swartz, Pediatrics, Women and Childrens Hospital of Buffalo, University at Buffalo, The State University of New York, Buffalo, NY and Stelios T. Andreadis, Department of Chemical and Biological Engineering, University at Buffalo, the State University of New York, Amherst, NY

Tissue engineered vascular constructs are promising for next generation vascular substitutes. Currently, one major challenge facing the field is developing biomimetic environments capable of providing appropriate biochemical and biomechanical signals to cells within bioengineered vascular grafts. Previously, we showed that TGF-β1 that was conjugated within fibrin hydrogels through the action of factor XIII prolonged phosphorylation of Smad2 and enhanced contractility of vascular grafts as compared to soluble growth factor. Vascular contractility increased significantly with the concentration of immobilized TGF-β1 in a biphasic manner. Here we examined the hypothesis that mechanical force could further enhance the function of TGF-b1 decorated vascular grafts. To this end, we developed a high throughput biomechanical plate to investigate the effects of different mechano-stimulation protocols on control or TGF-b1 decorated tissue engineered vascular constructs (TEVs). We found that mechanical stimulation induced circumferential alignment of MSCs, increased the collagen content and the Young's modulus of TEVs. Notably, pulsation at period of 2s and 5% distention significantly increased vascular reactivity through receptor and non-receptor mediated pathways suggesting increased myogenic function. Interestingly, mechanical stimulation of TGF-β1 conjugated TEVs further increased the vasoreactivity and mechanical properties. These results suggest that the optimal combination of mechanical stimulation and growth factor delivery yields more functional and mechanically robust vascular constructs.

(726f) Sustained Release Systems to Locally Expand Regulatory T Cell Populations and Suppress Inflammation

Stephen C. Balmert[1,2], Siddharth Jhunjhunwala[1,2], Giorgio Raimondi[3,4], John R. Vu[5], Louis D. Falo[2,5], Angus W Thomson[3,4,6] and Steven R. Little[1,2,6,7], (1)Bioengineering, University of Pittsburgh, Pittsburgh, PA, (2)McGowan Institute for Regenerative Medicine, Pittsburgh, PA, (3)Surgery, University of Pittsburgh, Pittsburgh, PA, (4)Starzl Transplantation Institute, Pittsburgh, PA, (5)Dermatology, University of Pittsburgh, Pittsburgh, PA, (6)Immunology, University of Pittsburgh, Pittsburgh, PA, (7)Chemical Engineering, University of Pittsburgh, Pittsburgh, PA

Aberrant inflammation leading to tissue destruction is responsible for transplant rejection, autoimmunity, and inflammatory diseases. Traditional therapies involve harsh regimens of systemic immunosuppression, which leave patients vulnerable to opportunistic infections. Consequently, there is considerable interest in developing novel methods to localize suppression to inflamed tissue.[1] One such therapeutic approach involves increasing the presence of suppressive regulatory T cells (Treg) at local sites,[2] which typically requires systemic infusion of *ex vivo* cultured cells.[3] As inspiration for an alternative method to increase Treg locally, we consider a naturally occurring mechanism employed by the immune system. Specifically, tolerogenic dendritic cells induce the differentiation of naive T cells to Treg by establishing a local immunosuppressive milieu, comprised of secreted cytokines (IL-2 and TGF-β).[4] In an inflammatory microenvironment, these factors may be insufficient to prevent differentiation of naive T cells to pro-inflammatory effector T cells; however, the immunosuppressant drug rapamycin (rapa) has been shown to preferentially suppress generation and proliferation of effector T cells, but not Treg.[5] With this rationale, we have developed and tested biomimetic sustained release microparticle (MP) formulations of various factors, including IL-2, TGF-β, and rapa (collectively FactorMP), which can dramatically increase populations of Treg *in vitro* and suppress local inflammation in a murine model of contact hypersensitivity.

To fabricate sustained release cytokine/drug formulations, IL-2, TGF-β, and rapamycin (rapa) were encapsulated in poly(lactic-co-glycolic acid) (PLGA) using emulsion-evaporation techniques, and the resulting FactorMP were characterized, as described elsewhere.[6] Empty PLGA MP (BlankMP) were used as controls. For *in vitro* Treg induction assays, na•ve CD4 T cells were cultured in the presence of the soluble or encapsulated cytokines and drugs. At day 4, cell phenotypes were analyzed by flow cytometry, and suppressive function was assessed by a standard mixed leukocyte reaction. For the contact hypersensitivity suppression study, either FactorMP or BlankMP were injected into the footpads of mice at days -11 and -6. On day -5, mice were sensitized with dinitrofluorobenzene (DNFB). A contact hypersensitivity response was elicited on day 1 by painting one footpad with DNFB, and measurements of footpad thickness (indicative of inflammation) were taken at 24, 48, and 72 hours post-elicitation. Subsequently, draining lymph nodes were isolated and T cell populations analyzed by flow cytometry.

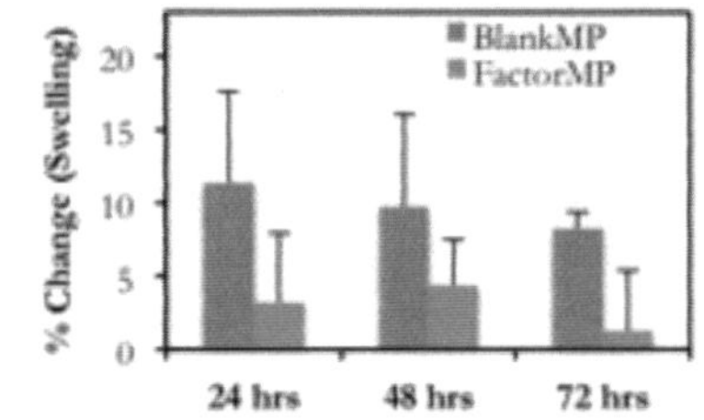

Figure 1. Percent increase in footpad thickness (swelling from inflammation) 24, 48, & 72 hrs post-elicitation with DNFB. Values reported relative to thickness of feet not painted with DNFB. Bars represent mean ± SE (n≥4).

IL-2, TGF-β, and rapa were successfully encapsulated in PLGA MP, with sizes ranging from 16.7±6.3 to 25.5±7.5μm (large enough to remain at an injection site). Each MP formulation provided sustained release of the encapsulated factor for at least 3 weeks *in vitro*, creating a sustained immunosuppressive, local microenvironment. Similar *in vitro* Treg induction efficiencies were observed for naive CD4 T cells cultured in the presence of FactorMP or soluble factors. Notably, the presence of soluble rapa or rapaMP, significantly ($p<0.05$) enhanced induction efficiency from 57±18% (with IL-2 and TGF-β alone) to 71±15%. Importantly, induced Treg expressed canonical Treg surface markers (CD25 and GITR), as well as FoxP3, a characteristic Treg transcription factor. Furthermore, FactorMP-induced Treg were functionally suppressive in a mixed leukocyte reaction. Finally, data from a contact hypersensitivity model (Figure 1) suggests that establishment of a local immunosuppressive milieu by FactorMP can decrease footpad swelling (local inflammation), presumably by increasing local Treg numbers.

In conclusion, we have successfully fabricated MP that provide sustained release of Treg-inducing factors. These FactorMP significantly increase numbers of functional Treg *in vitro*. Preliminary results also suggest that they can suppress inflammation *in vivo*. Ultimately, this FactorMP system may have the potential to treat a variety of inflammatory and autoimmune diseases, as well as transplant rejection.

[1] Thomson AW (ed). *Immunology and Medicine: Therapeutic Immunosuppression* (2001) Springer; 29:1-512.

[2] Sakaguchi S et al. *Immunol Rev* (2001) 182:18-32.

[3] Raimondi G et al. *J Immunol* (2010) 184:624-36.

[4] Horwitz DA et al. *Eur J Immunol* (2008) 38:901-937.

[5] Battaglia M et al. *Blood* (2005) 105:4743-4748.

[6] Jhunjhunwala S et al. *J Control Release* (2012) 159:78-84.

(726e) Wnt5a Conjugated Poly(ethylene glycol) - Gelatin Composite for Vascularized Tissue Engineering

Alpesh Patel[1,2], Akhilesh K. Gaharwar[3,4], Pinar Zorlutuna[1,2], Elif Karaca[2], Lina Schukar[2] and **Ali Khademhosseini**[1,2,4], (1)Harvard-MIT Division of Health Sciences and Technology, Massachusetts Institute of Technology, Cambridge, MA, (2)Center for Biomedical Engineering, Brigham and Women's Hospital, Harvard medical School,, Boston, MA, (3)David H. Koch Institute for Integrative Cancer Research, Massachusetts Institute of Technology, Cambridge, MA, (4)Wyss Institute for Biologically Inspired Engineering, Harvard University, Boston, MA

Embryonic stem cell (ESC) differentiation holds enormous potential for generating a renewable source of cells for regenerative medicine. However, directed in vitro differentiation of ESC has proven to be a challenging task. In early embryonic development, a tightly controlled series of cell-cell, cell-soluble factor and cell-matrix interactions create a microenvironment that provides a coordinated set of cues to developing cells. Wnt signaling molecules are among the most important cues that control key developmental events. Recently we have used microscale technologies to demonstrate the role of noncanonical Wnt11 in cardiogenesis and Wnt5a in directing endothelial cell differentiation of ESCs. Here, we propose to fabricate Wnt5a conjugated poly(ethylene glycol) vinyl sulfone (PEGVS)-gelatin composite to engineer vascular tissues by controlling ESCs differentiation. We fabricated photo cross-linked PEGVS-gelatin hydrogels with tunable mechanical and degradation properties by controlling the amount of gelatin. We then conjugated Wnt5A protein to 4-arm PEGVS via Michael-type addition utilizing the cysteine groups of the Wnt5a protein and analyzed the conjugation using matrix-assisted laser desorption/ionization (MALDI). The results indicated that one or two Wnt5a molecules were conjugated on the PEGVS arms. We further confirmed the conjugation by labeling the Wnt5a with a florescent molecule and observing the release kinetics. The biofunctionality of the Wnt5a conjugated hydrogels was evaluated by monitoring protein and gene expression of the encapsulated EBs. The immunocytochemistry results indicated that conjugation of Wnt5a to copolymer network results in higher expression of CD31 and higher sprouting compared to the control (without Wnt5a). Additionally, it was observed that CD31 was evenly distributed at the cell-cell junction and the expression pattern was similar to the fully mature endothelial cells. Furthermore, the endothelial differentiated of EBs was determined by monitoring the gene expression using quantitative polymerase chain reaction. The results indicated that conjugation of Wnt5a to PEGVS-Gelatin hydrogels up-regulated Wnt5a and CD31 gene expression compared to the control. Overall, we were able to successfully design Wnt5a conjugated PEGVS-gelatin hydrogels and able to modulate ESC signaling microenvironment based on noncanonical Wnt5a pathway for potential application in vascularized tissue engineering.

(776a) Isolation of Rare Circulating Tumor Cells and in Situ Culturing

Zhuo Zhang, Meggie M G Grafton and Sunitha Nagrath, Department of Chemical Engineering, University of Michigan, Ann Arbor, MI

Circulating tumor cells (CTCs) shed by primary and metastatic tumors are shown to carry important genetic and phenotypic information about tumor progression. Since late 1990s, massive microfluidic research effort has been devoted to isolate and characterize CTCs as a non-invasive method for cancer diagnosis and prognosis. One of the most effective methods to separate CTCs from blood is the immunoaffinity based CTC-chip with high capture efficiency and purity. However, the ability to use the captured CTCs to perform cellular characterization and drug screening is limited due to the fact that CTCs are rare. In order to overcome this challenge, we present a novel method to capture cancer cells from blood and conduct in situ cell culturing in a microfluidic platform. Low numbers of cancer cells spiked into whole blood are captured in a CTC-chip. Consequently, the captured cells are cultured on the same device. The growth of the cells is monitored and recorded for 72hrs using an on-stage culture chamber. This CTC isolation and culturing system can test the sensitivity of drugs and assist in therapy determination therefore extend the diagnostic potential of current CTC isolation technologies.

(776b) Engineered Microenvironments to Analyze Host-Tumor Cell Interactions

Eline Boghaert[1], Jason P. Gleghorn[1], KangAe Lee[1], Derek C. Radisky[2] and Celeste M. Nelson[1], (1)Princeton University, Princeton, NJ, (2)Mayo Clinic Cancer Center, Jacksonville, FL

The development of normal tissues, both natural and engineered, requires concerted interactions between cells and their local microenvironment. Similarly, breast tumor development is modulated by cues from the surrounding microenvironment, including interactions with neighboring non-tumor cells as well as the extracellular matrix (ECM). In addition to chemical signals, studies using homogeneous populations of breast cancer cell lines cultured in three-dimensional (3D) ECM have shown that the physical properties of the microenvironment, namely increased ECM stiffness, can stimulate the invasion of tumor cells. However, at early stages of breast cancer development, malignant cells are surrounded by normal epithelial cells, which have been shown to have a tumor-suppressive effect on co-cultured cancer cells. Here we used tissue engineering strategies to explore how the biophysical characteristics of the host microenvironment affect the proliferative and invasive tumor phenotype of the earliest stages of tumor development. We used a 3D microfabrication-based approach to engineer ducts comprised of normal mammary epithelial cells that contained a single tumor cell. We found that tumor cell phenotype was dictated by position in the duct, as tumor cell proliferation and invasion were enhanced at the ends, while these effects were blocked when the tumor cell was located elsewhere. Regions of invasion correspond with regions of high endogenous mechanical stress, as shown by finite element modeling and bead displacement experiments, and modulating the contractility of the host epithelium controlled the subsequent invasion of tumor cells. Combining micro-computed tomographic analysis with finite element modeling confirmed that regions of high mechanical stress correspond with regions of tumor formation *in vivo*. This work suggests that the mechanical tone of the non-tumorigenic host epithelium directs the phenotype of tumor cells, and provides additional insight into the instructive role of the mechanical tumor microenvironment. Our study also suggests that tissue engineering strategies can be used to unveil novel aspects of tumor development.

(776c) Elucidation of Mast Cell Localization Using a Microfluidic Device That Generates a Controllable Diffusion-Driven SCF Gradient

Meghaan M. Smith, Amir Shamloo, Maheswaran Mani, Milan Manchandia, Kenneth Weinberg and Sarah C. Heilshorn, Stanford University, Stanford, CA

Mast cells, granular cells that reside in tissue, have become a well-known modulator of wound healing. Investigators have proven that the absence of mast cells greatly impairs healing of skin wounds created in mast cell-deficient murine models. Subsequent reconstitution with bone marrow-derived mast cells (BMMCs) reversed this effect. These studies also demonstrated that the extent of mast cell degranulation, which promotes healing by increasing vascular permeability, neutrophil recruitment, and wound closure, correlates inversely with distance from the lesion. Thus, regulation of mast cell migration and localization is critical for wound healing.

In addition to its chemotactic properties, stem cell factor (SCF, aka kit ligand) is required for mast cell growth, differentiation, and survival; thus, SCF is key in regulating mast cell localization. To shed light on the mechanism behind SCF-dependent localization, we developed a gradient-generating microfluidic device, which offers several advantages over traditional chemotaxis assays, including establishment of a temporally stable gradient, minimal use of expensive reagents and cells, and the ability to directly visualize the migration of single cells. The devices employed for these studies generated a stable linear SCF gradient across a central channel containing BMMCs without exposing the cells to convective flow.The SCF concentration profile is dictated by Fickian diffusion across the cell culture chamber, which enables direct control over the gradient steepness and average concentration. A concentration gradient of 8 ng/ml/mm was determined to be sufficient to induce BMMC chemotaxis. Analysis of the time-lapse microscopic videos enabled tracking of the migratory paths of hundreds of individual cells, which were then quantified to determine speed, persistence time, and chemotactic index (CI) as a function of SCF gradient and concentration. Interestingly, we found that within the same 8 ng/ml/mm gradient, BMMCs exposed to the highest average SCF concentration (10 ng/ml) displayed greater migration efficiency (CI = 0.32), than cells exposed to lower concentrations (6 ng/ml, CI = 0.10, $p<0.05$), while cell speed was largely unaffected. As expected, BMMCs not subjected to an SCF gradient displayed random migration within the devices. These data reinforce the concept that chemokinesis, i.e. the effect of a chemokine on random cell motility and speed, is a distinct process from chemotaxis, which requires asymmetric reorganization of the cytoskeleton. These results further suggest that efficient cell polarization requires both a minimum gradient steepness as well as a minimum SCF concentration. Further studies of BMMC response to SCF may help elucidate the mechanism behind mast cell migration and localization in response to a lesion. The knowledge acquired through these studies will aid in the development of chemokine-delivery therapies for dermal wound healing of chronic lesions.

(776d) Mechanical Stretching Induced Mesenchymal Stem Cell Orientation

Chun Liu[1], Seungik Baek[2] and Christina Chan[1], (1)Chemical Engineering and Materials Science, Michigan State University, East Lansing, MI, (2)Mechanical Engineering, Michigan State University, East Lansing, MI

Tissues and cells in the body are exposed to a complex mechanical microenvironment. Therefore, mechanical stimulation is important for cell proliferation and differentiation. Recent interest has grown on the impact of mechanical stimulation, such as stiffness, surface topography, and mechanical stretching, on cell fate decision. In this study, we created an anisotropic mechanical environment by stretching the substrate membrane in an uniaxial direction, and recorded mesenchymal stem cell orientation response accordingly.

We investigated the alignment behavior of cells on an anisotropic surface. The experimental results agreed with our "cell active sensing" model, and suggest that the cells can sense and respond to the surface anisotropy by orienting in the direction of maximal effective stiffness. We apply this model to a uniaxial pre-stretched surface to predict cell orientation. After stretching a substrate unidirectionally, we applied a thin poly-L-lysine coating (~0.8nm) and culture the MSCs, whereby the cells do not experience the actual stretching. By culturing cells on a pre-stretched PDMS substrate membrane, the effective stiffness the cells sense in the stretched direction is larger than in the unstretched direction. This resulted in the cells aligning in the stretched direction, and contributed to the regulation of the cell focal adhesion, and further modulated cell signaling and differentiation.

Currently, most of the prior studies that focused on mechanical stress attempted to measure the cytoskeletal changes after active stretching, whereby the cells are seeded prior to applying the external force or stretch. However, in our experiment we seed the cells after the membrane is stretched, such that the cells do not experience the stretch due to external changes in force. Nevertheless the cells are able to sense the differential effective stiffness in the two directions and changed their cytoskeleton accordingly to orient in the direction of maximum stiffness. These results support our model of "active mechanosensing".

(776e) Controlling Embryonic Cell Sheet Migration Using Microfluidics

Melis Hazar[1], YongTae Kim[2], Jiho Song[1], Philip R. LeDuc[1], William Messner[1] and Lance A Davidson[3], (1)Mechanical Engineering, Carnegie Mellon University, Pittsburgh, PA, (2)Koch Institute for Integrative Cancer Research Massachsetts Institute of Technology, Cambridge, MA, (3)Bioengineering, University of Pittsburgh, Pittsburgh, PA

Embryonic development consists of a complex series of cell signaling, cell migration and cell differentiation events whereas morphogenesis is the process that controls the organized spatiotemporal distributions of these events. Cell sheet migration is central to embryonic development, to homeostasis of complex organs, and to disease pathology therefore studying cell migration within sheets is important. Embryos from the African Claw-toed frog, *Xenopus laevis*, are used to elucidate genes important in moving cells. However, little is known about the underlying mechanism by which cells in epithelial sheets coordinate their responses to growth factors, directional signals, and motility cues to direct sheet movement *in vivo*. One reason for this knowledge gap is the lack of technologies to control of the chemical microenvironment surrounding multicellular tissues. Here, we manipulate the chemical microenvironment with precise spatiotemporal delivery of cytoskeletal inhibitors and contraction-stimulating compounds using laminar flow interfaces with microfluidics. We find that microfluidics can pattern cell responses and control cell motility of multicellular sheets and may allow engineers the ability to initiate and control the outcome of synthetic morphogenetic programs. We use this approach to study motility of the cells within the sheets to these localized environments and understand how their coordinated behavior works with spatiotemporal control.

Microfluidics provides an opportunity to control chemical stimulation of biological systems using laminar flow interfaces with well-known flow modulation methods. To deliver precise chemical stimulation to a multicellular tissue, we have developed a system for controlling the inlet pressures to a microfluidic device by modulating fluidic resistance and capacitance. We employed this system to deliver chemicals over tissue explants from *Xenopus* embryos with a spatiotemporal control to study mechanical patterning and local control of cell sheet migration during epibolic-type morphogenetic movements. This approach enabled spatiotemporally controlled inhibition of cell sheet migration in an embryonic tissue to investigate the dynamic responses to localized inputs.

We observed that applying blebbistatin (BBS), a myosin inhibitor, to half of the animal cap explant, we could affect cell migration. When we stimulated the animal cap explant locally with BBS locally with our microfluidic system *in vivo*, sheet migration altered during epibolic-type morphogenetic movements. Using time-lapse microscopy, we were able to observe localized response, analyze the dynamics with kymographs, and construct spatiotemporal strain maps.

The development of new technologies to control the form of growing multicellular tissues advance the goals of regeneration and tissue engineering and produce new applications in bioengineering and medicine. We believe that patterning cell mechanics and controlling cell motility provide a means to initiate synthetic morphogenetic programs. In addition, the ability to control the form of multicellular tissues potentially has high impact in tissue engineering and regeneration applications in bioengineering and medicine.

(776f) Mechano-Transduction Pathway Interference with BMP-2 Signaling Cascade

Laure Fourel[1], **Jorge Almodovar**[1], Corinne Albiges-Rizo[2] and Catherine Picart[1], (1)Grenoble Institute of Technology, Grenoble, France, (2)Institut Albert Bonniot, Grenoble, France

The bone morphogenetic proteins (BMP) growth factors are essential signaling molecules in embryogenesis, tissue regeneration or cancer metastasis. Until now, the cellular effects of these proteins have mostly been investigated by adding them in solution in the cell culture medium even though growth factors are naturally not free to diffuse and are rather bound to ECM components including proteins and polysaccharides. In this work, we design a new biomimetic nanoassembly based on polyelectrolyte multilayer film made of a polypeptide and of hyaluronan which combines both matrix-bound presentation of growth factor bone morphogenetic protein 2 (BMP2) and modulation of the material's mechanical properties. We investigated the interplay between the stiffness of a biomaterial and the matrix-bound presentation of the growth factor BMP2 and its impact on cellular processes. Whereas C2C12 myoblast cells are round and poorly spread on soft films, cells displayed an increase of spreading and migration. By using a combined set of techniques including SiRNA strategy and pharmacological approach, we study how the mechano-transduction pathway can potentially interfere with the BMP-2 signalling cascade. We demonstrated a crosstalk between specific integrin and BMP2 receptors. Our results pointed out that signaling can be enhanced by the spatial association of integrin and BMP2 receptors.

(776g) Monocyte Chemoattractant Protein-1 Static Concentration Gradient in a 3D Collagen Matrix and Its Haptotactic Effect On Monocyte Migration

Neda Ghousifam and Heather Gappa Fahlenkamp, School of Chemical Engineering, Oklahoma State University, Stillwater, OK

Atherosclerosis is known as an inflammatory disease which is initiated by the accumulation of lipid substances in the subendothelial layer of major arteries, followed by adhesion of monocytes and lymphocytes to endothelial cells and their subsequent migration across the endothelial layer to the extracellular matrix (ECM). This adhesion and transmigration involves several steps, which are mediated by bioactive molecules named chemotactic cytokines (chemokines). Monocyte Chemoattractant Protein-1 (MCP-1) is one chemokine that plays an important role in monocyte trafficking across the endothelial layer. Many studies investigate the effect of soluble concentration gradients of chemokines, such as MCP-1, on monocyte migration by using a 2D experimental system. Studies have shown that soluble concentration gradients of chemokines drive, and possibly control cell migration. Recently, it was hypothesized that MCP-1 can bind with collagen, which is a major protein in the extracellular matrix (ECM). A 2D system may not adequately predict *in vivo* cell behavior due to the lack of the third dimension that is important for the creation of MCP-1 binding reaction in the ECM. A 3D tissue model consisting of a collagen matrix would be a better experimental model for studying the formation of MCP-1 static concentration gradient within the ECM. The goal of this research is to confirm the formation of an MCP-1 static concentration gradient within a simplified 3D vascular tissue model and examine its haptotactic effect on monocyte migration. The experimental results show that the 3D tissue model pre-treated with MCP-1 had a significant increase in the number of migrated monocytes in comparison with the control group lacking the pre-treatment. Also, when an MCP-1 antibody was used to block the MCP-1 within the collagen matrix, there was no significant difference between the number of migrated monocytes compared to the control group that was not treated with MCP-1. Furthermore, it is shown that the haptotactic (static) gradient of MCP-1 had the same effect and potency on monocyte migration as the chemotactic (soluble + static) gradient of MCP-1. The results suggest that the effect of the static MCP-1 concentration gradient on monocyte transmigration is stronger than the soluble MCP-1 concentration gradient in the collagen matrix. The results of this research will provide new information about the relationship between MCP-1 concentration gradients and monocyte transendothelial migration, and will lead to the development of an improved model to study transendothelial monocyte migration associated with inflammation.

(776h) STEP Enabled Long Time Culture of Primary Hepatocytes in Multiple Layers

Kevin Sheets, School of Biomedical Engineering and Sciences, Virginia Tech, Blacksburg, VA, Ji Wang, Engineering Science and Mechanics, Virginia Tech, Blacksburg, VA and Amrinder S. Nain, Mechanical Engineering, Virginia Tech, Blacksburg, VA

Introduction: Chronic diseases of the liver including hepatitis and cirrhosis account for a large portion of the 16,000 – 17,000 person-long waiting list for a liver transplantation. Only 5,000-6,000 liver donations are available each year, meaning only 63% of people needing a transplant are unable to receive one. To alleviate this deficit, bioartifical liver assist devices (BLADs) have been developed which perform the functions of a normal liver outside of the body. The design and implementation of more successful assist devices is driven by the need to maintain hepatocyte functionality over longer periods of time. In this work, hepatocytes are seeded onto an ECM-mimicking, highly aligned nanofibrous network. As single cells attach and spread, they pull against these fibers and cause deflections which allow migratory forces to be probed. We believe that an understanding of the magnitude and nature of these forces will allow improvements to be made to BLAD construction to create more viable assist devices for the many in need of them.

Figure 1. Hepatic arrangement around STEP Scaffolds

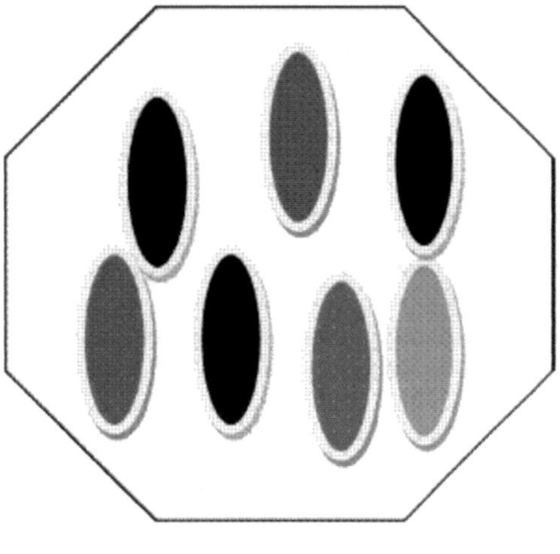

Figure 1. Hepatic arrangement around STEP Scaffolds

Materials and Methods: Using the previously-reported Spinneret-based Tunable Engineered Parameters (STEP) technique, nanofibrous scaffolds were created with fused intersections. Using xylene as a solvent, suspended polystyrene and polyurethane fibers of 500 nm and 300 nm diameter, respectively, were spun in criss-cross layers. Primary rat hepatocytes were seeded onto these constructs, which were then imaged with an incubating microscope. Migration behavior was analyzed via timelapse videos and fluorescence imaging. Cells were fixed and stained for nucleus (DAPI), actin (phalloidin) and focal adhesion (paxillin) locations.

Results and Discussion: After examining over 20 cells who had deflected nearby fibers, it was found that the average hepatocyte generated a force of approximately 1.5 nN.

Conclusions: We have shown that nanofibrous scaffolds carefully arranged for fiber spacing and layouts can cause cells to preferentially attach in specific shapes. According to focal adhesion immunohistochemistry, these configurations inherently display different attachment and migration forces which may prove useful in directed differentiation in the future.

AUTHOR INDEX

AUTHOR INDEX

AUTHOR INDEX

AUTHOR INDEX

AUTHOR INDEX